ENZYME REACTIONS and ENZYME SYSTEMS

ENZYMOLOGY

A Series of Textbooks and Monographs

Volume 1. Immobilized Enzymes, Antigens, Antibodies, and Peptides: Preparation and Characterization, *edited by Howard H. Weetall*

Volume 2. Subunit Enzymes: Biochemistry and Functions, *edited by Kurt E. Ebner*

Volume 3. Enzyme Structure and Function, *by S. Blackburn*

Volume 4. Enzyme Reactions and Enzyme Systems, *by Charles Walter*

Additional Volumes in Preparation

ENZYME REACTIONS and ENZYME SYSTEMS

Charles Walter
Department of Chemical Engineering
Cullen College of Engineering
University of Houston
Houston, Texas

MARCEL DEKKER, INC. New York and Basel

MARCEL DEKKER, INC.
270 Madison Avenue, New York, New York 10016

LIBRARY OF CONGRESS CATALOG CARD NUMBER: 75-22780
ISBN: 0-8247-6299-1
Current printing (last digit):
10 9 8 7 6 5 4 3 2 1

PRINTED IN THE UNITED STATES OF AMERICA

PREFACE

The purpose of this book is to describe how the catalytic properties of enzymes and multienzyme systems can be translated into mathematical models, and how these models, when compared with experimental observations, can be used to deduce information about chemical mechanisms of enzyme catalysis.

Enzyme kinetics is a tool for studying enzyme mechanisms. I emphasize the word "tool" because enzyme kinetics is just that. Individuals who would make enzyme kinetics an end in itself discover that there is no end to the number or complexity of mathematical models that one can concoct, nor to the amount of irrelevant data one can collect. Prominent theoretical kineticists have reffered to steady-state enzyme kinetics as that maze of enzyme algebra worth about two cents per pound. On the other hand, some individuals who use enzyme kinetics in laboratory experiments claim that such kinetic data is worth about one cent per pound, a value which is only a very small percent of what it costs to obtain.

Therefore, before beginning this book, it is appropriate to ask if theoretical and experimental studies of enzyme-catalyzed reactions are worthwhile. The answer, I believe, is that the construction of mathematical models for enzymes and multienzyme systems is useful, provided these models are based on real-life enzymes. Laboratory studies of the kinetics of isolated enzymes or of multienzyme systems can provide useful information in specific situations if the experimental design is carefully conceived. In other words, the usefullness of enzyme kinetics depends on successfully integrating data collection and model construction. Reasonable enzyme models come from data collected during innovative experiments, and innovated experiments are suggested by good models.

Previously enzyme kinetics has occasionally been an important part of studies on mechanisms of single, isolated enzymes, but more often it has been misused and/or overused. Thus, two cents per pound is not the intrinsic value of good enzyme kinetics, nor is one cent per pound the value of data collected during well-conceived experiments. These prices reflect the value of these items in the hands of enzymologists and other biochemists.

The value of enzyme kinetics can be improved by enriching the milieu in which it grows. This can be done by integrating the skills possessed by those who construct mathematical models of enzymes with those who conduct the experimental work.

Therefore, this book is not for "theoreticians." It is for practicing experimental enzymologists who wish to construct enzyme models worth more than two cents per pound and who wish to design kinetic experiments that will yield data worth what it costs to obtain. My job as author has been simplified considerably by the creation of certain computer programs which can be used to construct steady-state mathematical models for enzymes, to simulate complex chemical models of enzymes and multienzyme systems, to collect data during laboratory experiments with enzymes, and to estimate kinetic parameters for enzyme models from kinetic data. When understood and properly used, these programs can make a sophisticated scientist out of even the most stubborn "experimental" (i.e., antimathematics) enzymologist.

Chapters 1-5 deal with the kinetic properties of single enzymes in solution, with considerable emphasis on the assumptions that underlie steady-state enzyme models, the derivation of steady-state models with computers, enzyme model simulation on computers, and on-line data collection and analysis. I have also attempted to pose what I believe are some of the important questions for future work in this field. For example: How does one allocate error to enzyme kinetic models as well as to the experimental data? What is the best way to use experimental results to assist in the design of future experiments?

Chapters 6-8 discuss the kinetic properties of multienzyme systems, especially stability and control of far-from-equilibrium, sequential, catalyzed systems with typical biochemical feedback. As indicated in these chapters, mathematical models based on real-life, multienzyme systems together with key kinetic data collected from such systems can provide important information about the control and dynamics of complex metabolic systems.

Charles Walter

CONTENTS

CHAPTER 1

THE GENERAL THEORY OF ENZYME KINETICS

1.1 THE OVERALL CHEMICAL REACTION

A simple uncatalyzed chemical reaction may be thought of as a single overall reaction

$$S = P \tag{1.1}$$

or as a series of consecutive elementary reactions

$$S = \cdots = P \tag{1.2}$$

For most reactions Eq. (1.1) is a noncommital representation of the overall chemistry of a reaction of unknown mechanism, whereas Eqs. (1.2) are a representation inferring at least some mechanistic detail.

The *complete* set of elementary chemical reactions responsible for the overall process is the *reaction mechanism*. Enzymes that catalyze reactions of the type described by Eq. (1.1) function via at least three elementary steps:

$$\begin{aligned} &\text{STEP 1:} \quad E + S = X_1 \\ &\text{STEP 2:} \quad X_1 = \cdots = X_n \\ &\text{STEP 3:} \quad X_n = E + P \end{aligned} \tag{1.3}$$

Step 1, the "binding step," is the process responsible for increasing the probability of a close proximity of the substrate and the enzyme even though in biological systems these substances are ordinarily present in relatively small quantities. For enzymes that catalyze reactions involving more than one substrate,

$$\sum S_i = \sum P_i \tag{1.4}$$

The binding processes bring the various substrates into juxtaposition, and thus are also responsible for increasing the probability that the substrates will be in close proximity. Step 2, the "catalytic step(s)," are the processes responsible for the conversion of substrates to products and vice versa. In these steps the chemical events necessary for the overall reaction are completed; presumably the enzyme environment in the general vicinity of the binding sites is such that the probability of finding the substrate in a desired conformation is increased many fold. Since enzyme-substrate interactions involve multiple attachments between each substrate and the enzyme, highly specific substrate conformations can exist on the enzyme surface; in the case of multisubstrate reactions [Eq. (1.4)], multiple attachments can result in highly specific juxtapositions of the substrates as well. Such highly specific orientations would arise in single-point attachments or as a result of bimolecular collisions only infrequently. Finally, step 3, the "reverse binding step," is responsible for the desorption of the product from the enzyme surface and for the regeneration of the active form of the enzyme.

In the case of a specific enzyme-catalyzed reaction it may be difficult to distinguish each of the three processes illustrated in Eqs. (1.3). Some enzymes may not have an indigenous site for substrate binding; in these enzymes the substrate binding site is induced when the substrate is in the neighborhood of the enzyme or only after some other molecule has been bound to the enzyme. In the case of other enzymes there may be an indigenous binding site for only one portion of the substrate molecule; in these cases the binding sites for other points of attachment may be induced only when binding to the indigenous site occurs. The biological advantage of "protected" sites that are "exposed" only when they are needed is obvious. But in such cases it becomes difficult to distinguish the elementary chemical steps that contribute to binding or desorption from the catalytic steps.

Enzymes function by providing a suitable environment wherein an efficient pathway between the substrates and products can be formulated. This new pathway usually accounts for practically all of the conversion of substrates to products in biochemical pathways. As usual, it is assumed that each step involves an Eyring-type complex followed by the formation of thermodynamically stable enzyme-containing compounds (steps 1 and 2), or free enzyme and the product of the overall reaction (step 3). Presumably the largest activation energy for the catalytic steps is considerably lower than the activation energy for the uncatalyzed reaction. This concept is illustrated in Fig. 1.1. The broken line is the reaction

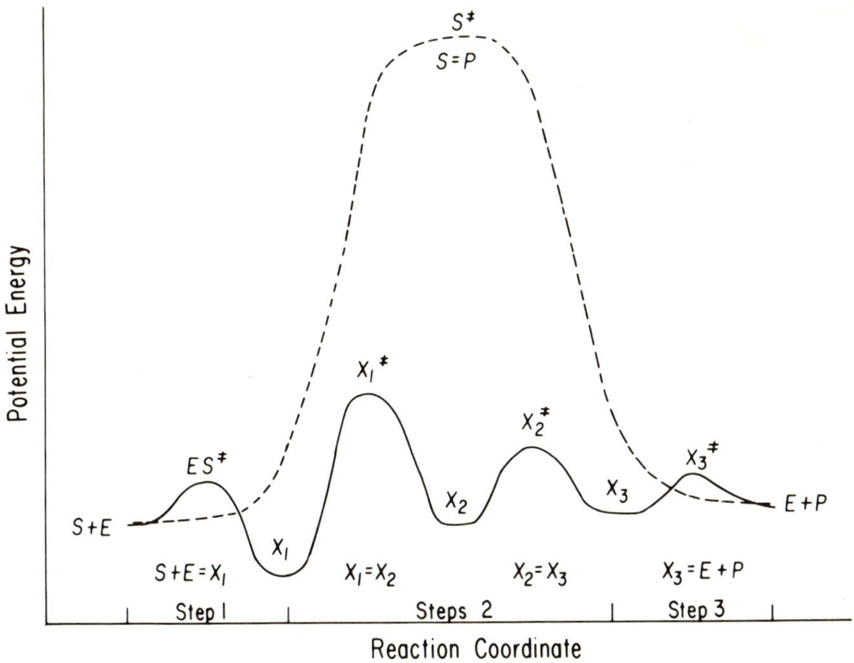

FIG. 1.1. Potential-energy reaction profile for an uncatalyzed (broken line) and an enzyme-catalyzed (solid line) reaction.

profile for an uncatalyzed reaction, and the solid line is the profile for an enzyme-catalyzed reaction involving one binding step, two catalytic steps, and one desorption step.

1.2 DEVELOPMENT OF EMPIRICAL KINETIC EQUATIONS FOR ENZYME-CATALYZED REACTIONS

Equations (1.4) describe the general mechanism for enzyme-catalyzed reactions involving one substrate and one product. The differential equations describing this mechanism are

$$\frac{dS}{dt} \equiv \dot{S} = -k_1 ES + k_{-1}X_1$$

$$\dot{X}_1 = -(k_2 + k_{-1})X_1 + k_1 ES + k_{-2}X_2$$

$$\vdots$$

$$\dot{X}_i = -(k_{i+1} + k_{-i})X_i + k_i X_{i-1} + k_{-(i+1)}X_{i+1}$$

$$[i=2, \ldots, n-1;\ n>2]$$

$$\vdots$$

$$\dot{X}_n = -(k_{n+1} + k_{-n})X_n + k_n X_{n-1} + k_{-(n+1)}EP$$

$$\dot{P} = k_{n+1}X_n - k_{-(n+1)}EP$$

$$-\dot{E} = -k_{-1}X_1 - k_{n+1}X_n + (k_1 S + k_{-(n+1)}P)E \tag{1.5}$$

where k_i and k_{-i} are the forward and reverse rate constants for each of the steps in Eqs. (1.4). For closed systems the conservation of mass requires that

$$E(0) = E + \sum_{i=1}^{n} X_i$$

$$S(0) = S + P + \sum_{i=1}^{n} X_i \tag{1.6}$$

where E(0) and S(0) are the total amounts of enzyme and substrate added initially.

The analytic solution for Eqs. (1.5) and (1.6) have not been obtained even for the trivial case n = 1. However, because in many in-vitro experimental situations

$$S(0) \gg E(0) \tag{1.7}$$

it is often possible to obtain an *approximate* solution over a short time interval, $t_1 \leq t \leq t_2$. If, soon after mixing $X_i(t)$ and $E(t)$ are sufficiently "well-behaved" functions that during this time interval it is permissible to treat

$$-k_{-(n+1)} E(t^*) = a(t^*) \tag{1.8}$$

and

$$k_{n+1} X_{n+1}(t^*) = b(t^*) \tag{1.9}$$

as constants, an approximate solution can be obtained for t* in the interval $t_1 \leq t \leq t_2$. The differential equation for the appearance of the product is

$$\dot{p} = a(t^*)\, p + b(t^*) \tag{1.10}$$

where the lower-case variables represent the chemical concentrations during the time interval when the approximations in Eqs. (1.8) and (1.9) are valid. Integration of Eq. (1.10) over the interval t_1 to t_2 leads to

$$p(t) = \left[p(t_1) + \frac{b(t^*)}{a(t^*)}\right] \exp\,[a(t^*)(t - t_1)] - \frac{b(t^*)}{a(t^*)} \tag{1.11}$$

For many enzymes Eq. (1.11) is an adequate description of short segments of the in-vitro time course of the catalyzed reaction when t_1 is as small as a few milliseconds or as large as desired. Eq. (1.11) does not apply during the interval when the reactants are mixing, because it is during this transient phase that the $X_i(t)$ and $E(t)$ are rapidly changing functions. Furthermore, Eq. (1.11) could not be used if $X_i(t)$ and $E(t)$ were periodic functions. Fortunately, for most in-vitro experiments with enzymes in closed systems, $X_i(t)$ and $E(t)$ can be treated as quasiconstant parameters for intervals of up to several seconds or longer. However, since periodic functions of concentrations are well known in controlled, open enzyme-catalyzed systems, extrapolation of this method to open systems is not always possible.

One consequence of inequality (1.7) is that insignificant quantities of the material originally introduced as substrate are bound to the enzyme at any given instant. Under these conditions the second equation in (1.6) can be written in the approximate form,

$$S(0) = S(t) + P(t) \tag{1.12}$$

for all t. Substitution from Eq. (1.12) into Eq. (1.11) provides the equation for s(t):

$$s(t) = S(0) - \left[p(t_1) + \frac{b(t^*)}{a(t^*)}\right] \exp\,[a(t^*)(t - t_1)] - \frac{b(t^*)}{a(t^*)} \tag{1.13}$$

If t_1 is sufficiently early during the overall reaction that

$$p(t_1) << - \frac{b(t^*)}{a(t^*)} \tag{1.14}$$

Eqs. (1.11) and (1.13) become

$$s(t) - S(0) = p(t) = \frac{b(t^*)}{a(t^*)} \{\exp\,[a(t^*)(t - t_1)] - 1\} \tag{1.15}$$

In the case of many enzymes Eq. (1.15) is an adequate description of the time course of the catalyzed reaction beginning a few milliseconds after mixing and lasting up to several seconds or longer.

If we differentiate Eq. (1.12), we obtain

$$\dot{S} = -\dot{P} \tag{1.16}$$

but if we differentiate Eq. (1.6), we obtain

$$\dot{S} = -\dot{P} - \sum_{i=1}^{n} \dot{X}_i = -\dot{P} + \dot{E} \tag{1.17}$$

Reconciliation of Eqs. (1.16) and (1.17) is possible only if $|\dot{E}|$ is small compared to $|\dot{P}|$ and $|\dot{S}|$. A common stragedy in enzyme kinetics is to suppose that since the coefficients in Eqs. (1.10) can be treated as constant during the short interval t_1 to t_2, $|\dot{e}|$ and $|\dot{x}_i|$ can be assumed small compared to $|\dot{s}|$ and $|\dot{p}|$. By using $\dot{e} = 0$ and $\dot{x}_i = 0$ explicitly, it is possible to generate approximate definitions for a(t*) and b(t*) for various n; these definitions depend on the individual rate constants, the total enzyme concentration, and S(t*) and P(t*):

$$a(t^*) = -\frac{k_{-(n+1)}\, E(0)}{1 + \dfrac{S(t^*)}{K_f} + \dfrac{P(t^*)}{K_r}} \tag{1.18}$$

$$b(t^*) = \frac{\dfrac{V_f}{K_f}\left[S(t^*) - \dfrac{P(t^*)}{{}^{n+1}K_0}\right] + k_{-(n+1)}\, E(0)\, P(t^*)}{1 + \dfrac{S(t^*)}{K_f} + \dfrac{P(t^*)}{K_r}} \tag{1.19}$$

where

$$V_f = \frac{E(0)}{\displaystyle\sum_{i=1}^{n} \sum_{s=i}^{n} \frac{{}^{s}K_i}{k_{s+1}}}$$

$$V_r = \frac{E(0)}{\displaystyle\sum_{i=1}^{n} \sum_{s=0}^{i-1} \frac{{}^{s}K_i}{k_{s+1}}}$$

$$K_f = \frac{\displaystyle\sum_{s=0}^{n} \frac{{}^{s}K_0}{k_{s+1}}}{\displaystyle\sum_{i=1}^{n} \sum_{s=i}^{n} \frac{{}^{s}K_i}{k_{s+1}}}$$

$$K_r = \frac{\displaystyle\sum_{s=0}^{n} \frac{{}^{s}K_0}{k_{s+1}}}{\left[\displaystyle\sum_{i=1}^{n} \sum_{s=0}^{i-1} \frac{{}^{s}K_i}{k_{s+1}}\right] {}^{n+1}K_0}$$

and

$${}^{n+1}K_0 = \prod_{r=1}^{n+1} \frac{k_{-r}}{k_r}$$

$${}^{s}K_i = \prod_{r=i+1}^{s} \frac{k_{-r}}{k_r}$$

$${}^{s}K_0 = \prod_{r=1}^{s} \frac{k_{-r}}{k_r}$$

If e(t) and x_i(t) do not change much compared to s(t) and p(t) during the interval t_1 to t_2, but during this same interval s(t) and p(t) do not change much compared to S(t) and P(t) during the interval $0 \le t \le \infty$, S(t*) and P(t*) in Eqs. (1.18) and (1.19) can be treated as constants. Clearly, the choice for the numerical value of S(t*) and P(t*) will depend upon t*; a weighted average value for s(t) and p(t) during the appropriate t_1 to t_2 interval should suffice to define a(t*) and b(t*).

If t_1 is chosen sufficiently early during the overall reaction,

$$S(0) >> P(t^*) << K_r \tag{1.20}$$

$${}^{n+1}K_0 S(0) >> P(t^*) << \frac{V_f}{E(0)} \frac{S(0)}{K_f} \frac{1}{k_{-(n+1)}} \tag{1.21}$$

Using the approximation

$$S(t^*) = S(0) \tag{1.22}$$

we obtain approximate values for a(t*) and b(t*) when t* is very small:

$$a(0) = -\frac{k_{-(n+1)}E(0)}{1 + \frac{S(0)}{K_f}} \tag{1.23}$$

$$b(0) = \frac{\frac{V_f}{K_f} S(0)}{1 + \frac{S(0)}{K_f}} \tag{1.24}$$

The definitions in Eqs. (1.23) and (1.24) can be used during the early stage of the overall reaction. For many enzymes this early stage begins a few tenths of a second after mixing, and it lasts several seconds or longer. For later stages of the overall reaction, the more general definitions provided in Eqs. (1.18) and (1.19) should be used.

1.3 THE RELATIONSHIP BETWEEN THE KINETIC EQUATIONS AND INITIAL RATE EQUATIONS

The usual procedure in chemical kinetic studies is to attempt to obtain expressions relating concentrations, time, and the parameters of a suspected mechanism. We have followed this procedure for a simple mechanism of enzyme catalysis in the preceeding section; the kinetic expressions we derived are approximate, but for a large number of enzyme-catalyzed reactions they are perfectly adequate. It is somewhat surprising therefore that even though these expressions are readily available, most "kinetic" studies involving enzymes do not even utilize time as a variable. The most common approach is to estimate the slope of P(t) or S(t) at a specific instant. Usually a time during the early portion of the overall reaction (but well after mixing is complete) is chosen for the estimate of $\dot{P}$. In such studies the initial conditions [E(0), S(0), and P(0)] are the "varied" quantities, and it is the relationship between $\dot{P}$ and these initial conditions that is used to investigate the "kinetic" behavior.

It is not easy to accept the traditional justifications for

discarding the wealth of information in P(t) and S(t). It is true that some enzymes are not stable during in-vitro catalysis for more than a few minutes. In these cases information about the catalytic properties of the native enzyme may be concentrated in the early portion of the overall reaction. More often, however, this information is distributed throughout the entire time course. In such cases the decision to use estimated early rates rather than all of p(t) or s(t) may not be justified.

When the experimental situation requires the use of estimated initial rates, the appropriate expression relating the rate, the initial conditions, and the parameters is

$$\dot{p}(0) \equiv v(0) = b(0) = \frac{V_f\ S(0)}{K_f + S(0)} \quad [P(0) = 0] \tag{1.25}$$

Equation (1.25) is the well-known expression first derived for initial quasi-steady-state rates by Briggs and Haldane[1] in 1925. If P(0) is not zero, the appropriate expression is

$$v(0) = \frac{V_f\ S(0)}{K_f + S(0) + \dfrac{K_f\ P(0)}{K_r}} \tag{1.26}$$

Equation (1.26) is the well-known expression relating initial quasi-steady-state rates and initial concentrations of substrate, product, and enzyme.

1.4 GENERAL INITIAL QUASI-STEADY-STATE RATE EQUATIONS FOR MULTISUBSTRATE REACTIONS

A majority of enzyme-catalyzed reactions involve at least two substrates and/or products. In such reactions it is possible for the substrates to bind and/or the products to desorb from the enzyme in a completely *random order*, in a *preferred sequence*, or in a completely obligatory *ordered sequence*. Furthermore, in the case of ordered sequence mechanisms there are two distinct subclasses of reactions: In *transaminase-type mechanisms* (sometimes referred to as "ping-pong" mechanisms), one or more of the products is desorbed from the enzyme before the last substrate combines with the enzyme; in *dehydrogenase-type mechanisms* (sometimes ambiguously referred to simply as "ordered" mechanisms), all substrates are bound to the enzyme before any of the products are desorbed.

There is a need for a simple, easy-to-use, but comprehensive nomenclature system for multisubstrate, multiproduct enzyme-catalyzed reactions. Here we shall refer to reactions as "nS, mP-random," "nS, mP-preferred," or "nS, mP-ordered" where n is the number of substrates and m is the number of products in the overall reaction, and the "-random," "-preferred," or "-ordered" suffix distinguishes among these three possibilities. In addition for nS, mP-ordered mechanisms with $n > 1$ and $m > 1$, the sequence will be annexed in parentheses to distinguish among the possible sequences. Thus for two-substrate, two-product, ordered-sequence mechanisms, "2S, 2P-ordered-(SSPP)" refers to the dehydrogenase-type reaction, but "2S, 2P-ordered-(SPSP)" refers to the transaminase-type mechanism. "3S, 2P-ordered-(SSSPP)" and "2S, 3P-ordered-(SSPPP)" refer to dehydrogenase-type reactions involving three substrates and two products, and two substrates and three products, respectively, but "3S, 2P-ordered-(SSPSP)" and "2S, 3P-ordered-(SPSPP)" refer to the transaminase-type mechanisms for these reactions. Note that the other possible sequences for the transaminase-type mechanisms, 3S, 2P-ordered-(SPSSP) and 2S, 3P-ordered-(SPPSP), are kinetically (but not chemically!) equivalent to 3S, 2P-ordered-(SSPSP) and 2S, 3P-ordered-(SPSPP), respectively. For reactions involving three substrates and three products, the possible ordered-sequence mechanisms are: 3S, 3P-ordered-(SSSPPP) (the dehydrogenase-type mechanism), and 3S, 3P-ordered-(SSPSPP) [which is equivalent to 3S, 3P-ordered-(SPPSSP)], 3S, 3P-ordered-(SPSSPP) [which is equivalent to 3S, 3P-ordered-(SSPPSP)], and 3S, 3P-ordered-(SPSPSP); these last five mechanisms are of the transaminase type. The nonequivalent four-substrate, four-product, ordered-sequence mechanisms are: 4S, 4P-ordered-(SSSSPPPP), -(SSSPSPPP), -(SSPSSPPP), -(SPSSSPPP), -(SSSPPSPP), -(SSPSPSPP), -(SPSSPSPP), -(SSPPSSPP), -(SPSPSSPP), and -(SPSPSPSP). It is obvious that this system of nomenclature can be extended *unambigu-*

ously to ordered-sequence mechanisms involving any number of substrates and products. No matter how complex the overall reaction is, the sequence will be immediately obvious from the designation.

The general mechanism for 2S, 2P-random, 2S, 2P-preferred, and 2S, 2P-ordered-(SSPP) is

$$\text{STEPS 1:}\quad S_1 + E \underset{k_{-1}}{\overset{k_1}{\rightleftharpoons}} X_1$$

$$S_2 + E \underset{h_{-1}}{\overset{h_1}{\rightleftharpoons}} X_2$$

$$S_2 + X_1 \underset{k_{-2}}{\overset{k_2}{\rightleftharpoons}} X_3$$

$$S_1 + X_2 \underset{h_{-2}}{\overset{h_2}{\rightleftharpoons}} X_3$$

$$\text{STEPS 2:}\quad X_3 \underset{k_{-3}}{\overset{k_3}{\rightleftharpoons}} \cdots \underset{k_{-(n-1)}}{\overset{k_{n-1}}{\rightleftharpoons}} X_n$$

$$\text{STEPS 3:}\quad X_n \underset{k_{-n}}{\overset{k_n}{\rightleftharpoons}} X_{n+1} + P_1$$

$$X_n \underset{h_{-n}}{\overset{h_n}{\rightleftharpoons}} X_{n+2} + P_2$$

$$X_{n+1} \underset{k_{-(n+1)}}{\overset{k_{n+1}}{\rightleftharpoons}} E + P_2$$

$$X_{n+2} \underset{h_{-(n+1)}}{\overset{h_{n+1}}{\rightleftharpoons}} E + P_1 \qquad (1.27)$$

As before, steps 1 are the binding steps, steps 2 are the catalytic steps, and steps 3 are the reverse binding steps. The definitions of the equilibrium constants are:

$$K_1 \equiv \frac{k_1}{k_{-1}} = \frac{X_1(\infty)}{E(\infty)S_1(\infty)} \qquad H_1 \equiv \frac{h_1}{h_{-1}} = \frac{X_2(\infty)}{E(\infty)S_2(\infty)}$$

$$K_2 \equiv \frac{k_2}{k_{-2}} = \frac{X_3(\infty)}{X_1(\infty)S_2(\infty)} \qquad H_2 \equiv \frac{h_2}{h_{-2}} = \frac{X_3(\infty)}{X_2(\infty)S_1(\infty)}$$

$$K_n \equiv \frac{k_n}{k_{-n}} = \frac{X_{n+1}(\infty)P_1(\infty)}{X_n(\infty)} \qquad H_n \equiv \frac{h_n}{h_{-n}} = \frac{X_{n+2}(\infty)P_2(\infty)}{X_n(\infty)}$$

$$K_{n+1} \equiv \frac{k_{n+1}}{k_{-(n+1)}} = \frac{E(\infty)P_2(\infty)}{X_{n+1}(\infty)} \qquad H_{n+1} \equiv \frac{h_{n+1}}{h_{-(n+1)}} = \frac{E(\infty)P_1(\infty)}{X_{n+2}(\infty)} \tag{1.28}$$

There are two redundancies implied by Eqs. (1.28):

$$\frac{k_1k_2}{k_{-1}k_{-2}} = K_1K_2 = H_1H_2 = \frac{h_1h_2}{h_{-1}h_{-2}} \tag{1.29}$$

$$\frac{k_nk_{n+1}}{k_{-n}k_{-(n+1)}} = K_nK_{n+1} = H_nH_{n+1} = \frac{h_nh_{n+1}}{h_{-n}h_{-(n+1)}} \tag{1.30}$$

Since Eqs. (1.29) and (1.30) are independent of time, they must be maintained whether or not the overall reaction is at theymodynamic equilibrium.

Random mechanisms are multiple-sequence pathways wherein all the paths are of approximately equal importance. For 2S, 2P-random mechanisms,

$$k_1 = h_1, \; k_{-1} = h_{-1}, \; k_2 = h_2, \text{ and } k_{-2} = h_{-2} \tag{1.31}$$

because the substrate binding is random, and

$$k_n = h_n, \; k_{-n} = h_{-n}, \; k_{n+1} = h_{n+1}, \text{ and } k_{-(n+1)} = h_{-(n+1)} \tag{1.32}$$

because the product binding is also random. When Eqs. (1.31) are approximately valid, S_1 and S_2 bind E with about equal facility, and S_1 combines with X_2 about as easily as does S_2 with X_1. When Eqs. (1.32) are approximately valid, P_1 and P_2 desorb from X_n with approximately equal facility, and P_1 desorbs from X_{n+2} about as easily as does P_2 from X_{n+1}.

Preferred-sequence mechanisms are multiple-sequence pathways wherein some of the paths are significantly favored over others. For 2S, 2P-preferred mechanisms,

$$k_1 > h_1,\ k_{-1} > h_{-1},\ k_2 > h_2,\ \text{and}\ \ k_{-2} > h_{-2} \tag{1.33}$$

implies that there is a preferential binding of the free enzyme with S_1, and that S_2 binds X_1 better than S_1 binds X_2. The distinction between the various substrates in random mechanisms is arbitrary, so one is free to assign subscripts to the S's in whatever manner one wishes. Normally in preferred- and ordered-sequence mechanisms S_1 is the substrate that combines best with the free enzyme, and in general S_i is the substrate that combines with the enzyme compound containing all the substrates with subscripts less than i.

In these same mechanisms,

$$k_n < h_n,\ k_{-n} < h_{-n},\ k_{n+1} < h_{n+1},\ \text{and}\ \ k_{-(n+1)} < h_{-(n+1)} \tag{1.34}$$

implies that there is a preferential desorption of P_1 from X_n, and that P_2 is formed preferentially via X_{n+1}. P_1 is usually defined as the product preferentially desorbed from X_n, and P_i as the product desorbed preferentially from X_{n-1+i}.

In 2S, 2P-ordered-(SSPP) mechanisms,

$$k_1 >> h_1,\ k_{-1} >> h_{-1},\ k_2 >> h_2,\ \text{and}\ \ k_{-2} >> h_{-2} \tag{1.35}$$

implies that the free enzyme combines only with S_1, and S_2 binds only with X_1, and

$$k_n \ll h_n,\ k_{-n} \ll h_{-n},\ k_{n+1} \ll h_{n+1},\ \text{and}\ \ k_{-(n+1)} \ll h_{-(n+1)} \tag{1.36}$$

implies that P_1 is formed only from X_n, and P_2 is desorbed exclusively from X_{n+1}. Equations (1.29) and (1.30) require that K_1K_2 equal H_1H_2, and K_nK_{n+1} equal H_nH_{n+1}. This requirement is not inconsistent with the inequalities in (1.35) and (1.36), which require only that *individual* rate constants become small.

Equation (1.25) describes the relationship between initial quasi-steady-state rates and initial substrate concentrations for single-substrate enzyme-catalyzed reactions, but *under certain conditions* an identical mathematical expression is applicable to multisubstrate mechanisms. For example, if the mechanism involves an obligatory order of substrate binding, a similar rectangular hyberbolic relationship can be derived for multisubstrate reactions:

$$v(0) = \frac{\alpha_1\ S(0)}{1 + \beta_1\ S(0)} \tag{1.37}$$

where S can represent any of the sequential substrates, and α_1 and β_1 depend on the initial concentrations of all the substrates but S. For ordered-sequence mechanisms the reciprocal of β_1 is usually referred to as the "Michaelis constant with respect to S", and α_1/β_1 as the "limiting velocity with respect to S."

For two-substrate mechanisms involving random or a preferential sequence of substrate binding, the relationship between the initial quasi-steady-state rate and the initial concentration of either substrate is

$$v(0) = \frac{\alpha_1\ S(0) + \alpha_2\ S(0)^2}{1 + \beta_1\ S(0) + \beta_2\ S(0)^2} \tag{1.38}$$

where the α's and β's depend on the initial concentration of the other substrate (i.e., the substrate not represented by S).

In general, for multisubstrate mechanisms involving random or preferred sequences of substrate binding, the relationship between the initial quasi-steady-state rate and the initial concentration of any of the substrates is

$$v(0) = \frac{\sum_{i=1}^{p} \alpha_i \, S(0)^i}{1 + \sum_{i=1}^{p} \beta_i \, S(0)^i} \tag{1.39}$$

where p is the total number of substrates in the overall reaction and the α's and β's depend on the initial concentrations of all the substrates except S.

CHAPTER 2

THE QUASI-STEADY-STATE ASSUMPTION

2.1 INTRODUCTION TO THE CONCEPT OF A QUASI-STEADY STATE IN A CLOSED SYSTEM

One can obtain an idea about the implications of using a quasi-steady-state approximation to derive expressions for enzyme mechanisms by considering the validity of the approximation when it is applied to the simplest enzyme-catalyzed reaction

$$E + S \underset{k_{-1}}{\overset{k_1}{\rightleftharpoons}} X_1 \underset{k_{-2}}{\overset{k_2}{\rightleftharpoons}} E + P \tag{2.1}$$

Admittedly, Eqs. (2.1) are probably not an adequate description of the mechanism for any real enzyme. If the quasi-steady-state approximation works for Eqs. (2.1), that is no guarantee that the approximation will work for a more complex mechanism. However, if the quasi-steady-state assumption should be a poor approximation for Eqs. (2.1), the implication is that it might also be a poor approximation for more complex enzyme mechanisms.

The differential equation for $\dot{X}_1$ in Eqs. (2.1) is

$$\dot{X}_1(t) = k_1\, E(t)\, S(t) + k_{-2}\, E(t)\, P(t) - (k_{-1} + k_2)\, X_1(t) \tag{2.2}$$

Since $X_1(0) = 0$, $X_1(t)$ grows from zero to $X_1(\infty)$, the eventual equilibrium concentration of X_1. If $P(0) = 0$, $\dot{X}_1(0) = k_1 E(0)\, S(0)$, but $\dot{X}_1(\infty) = 0$. On the other hand, $\dot{P}(0) = 0$ and $\dot{P}(\infty) = 0$, although $\dot{P}(t) > 0$ for all $0 < t < \infty$. The maximum value for $\dot{P}(t)$,

$$\dot{P}(t) = k_2 X_1(t) - k_{-2} E(t) P(t) \tag{2.3}$$

is $\dot{P}_m = k_2 E(0)$ because $k_{-2} E(t) P(t)$ is necessarily nonnegative and the maximum value possible for $X_1(t)$ is $E(0)$. Thus the ratio $\dot{P}_m/\dot{X}_1(0) = k_2/[k_1 S(0)]$. $S(0)$ is usually in the range 10^{-5} to 10^{-3} *M*, k_1 is usually in the range $10^8 - 10^9$/*M*·sec., and k_2 is usually in the range $10^3 - 10^4$/sec. Thus, the expected value for $\dot{P}(0)/\dot{X}_1(0)$ is zero, but a typical ratio for $\dot{P}_m/\dot{X}_1(0)$ is in the range 0.1. These facts, $\dot{P}(0) \ll \dot{X}_1(0)$ and $\dot{P}_m \simeq \dot{X}_1(0)$, must somehow be reconciled with the concept of an initial quasi-steady-state wherein *at very early times* $|\dot{X}_1|$ must be small compared to $|\dot{P}|$ and $|\dot{S}|$.

If $X_1(t)$ goes through a maximum at $t = t^*$, the steady-state assumption is literally valid at that instant. However, if $X_1(t)$ passes through a maximum, then $\ddot{X}_1(t^*) < 0$. Differentiation of Eq. (2.2) yields

$$\ddot{X}_1(t^*) = [k_1 \dot{S}(t^*) + k_{-2} \dot{P}(t^*)] E(t^*) + [k_1 S(t^*) + k_{-2} P(t^*)] \dot{E}(t^*) - (k_{-1} + k_2) \dot{X}_1(t^*) \tag{2.4}$$

Substituting $\dot{X}_1(t^*) = \dot{E}(t^*) = 0$ and $\dot{S}(t^*) = -\dot{P}(t^*)$, we obtain

$$\ddot{X}_1(t^*) = (k_{-2} - k_1) E(t^*) \dot{P}(t^*) \tag{2.5}$$

Since k_{-2}, k_1, $E(t^*)$, and $\dot{P}(t^*)$ are all nonnegative, $k_{-2} < k_1$ is necessary for $\ddot{X}_1(t^*) < 0$. In other words, k_{-2} must be less than k_1 for $X_1(t)$ to pass through a maximum.

Since $\dot{X}_1(0) > 0$, a maximum in $X_1(t)$ is a prerequisite for a minimum. Therefore, $X_1(t)$ cannot experience a minimum when $k_{-2} \geq k_1$. Moreover, for $k_{-2} < k_1$, a minimum requires that

$\dot{E}(t^{**}) = \dot{X}_1(t^{**}) = 0$ when $\ddot{X}_1(t^{**}) > 0$. But according to Eqs. (2.5) $\ddot{X}_1(t^{**}) > 0$ is not possible when $k_{-2} < k_1$. Thus, if $k_{-2} \geq k_1$, $X_1(t)$ increases monotonically toward $X_1(\infty)$, but if $k_{-2} < k_1$, $X_1(t)$ increases to a maximum concentration and then decreases monotonically toward $X_1(\infty)$. This behavior is illustrated in Fig. 2.1.

When $k_1 > k_{-2}$, $\dot{X}_1(t^*) = 0$, and the steady-state assumption is exact. Under these conditions we may substitute $\dot{X}_1(t^*) = 0$ into Eq. (2.2) and obtain

$$X_1(t^*) = 0.5(\eta - \{\eta^2 - 4E(0)[\eta - E(0)]\}^{1/2}) \tag{2.6}$$

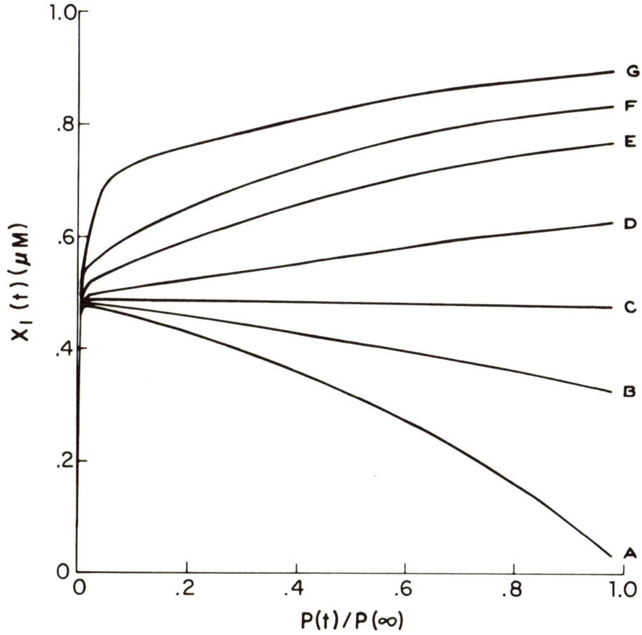

FIG. 2.1. $X_1(t)$ versus extent of overall reaction for various values of k_{-2}/k_1. For all curves $k_1 = 10/mM\cdot\tau$, $k_{-1} = 0.1/\tau$, $k_2 = 1/\tau$ where τ is the time unit for the rate constants, and $S(0) = 100\ E(0) = 0.1$ mM. The values of k_{-2}/k_1 are: curve A, 0; curve B, 0.5; curve C, 0.9; curve D, 2; curve E, 5; curve F, 10; curve G, 100.

where $\eta = E(0) + S(0) + K_f + (K_f - K_r)\ P(t^*)/K_r$, and $K_r < K_f$ have the meanings assigned after Eqs. (1.18) and (1.19) when n = 1. Substitution from Eq. (2.6) into Eq. (2.3) yields

$$\dot{P}(t^*) = \frac{k_2 + k_{-2}\ P(t^*)}{2} \cdot (\eta - \{\eta^2 - 4E(0)[\eta - E(0)]\}^{\frac{1}{2}}) - k_{-2}\ E(0)\ P(t^*) \quad (2.7)$$

When $k_1 \leq k_{-2}$, $X_1(t) > 0$ for all $0 < t < \infty$, and the steady-state assumption cannot be exactly correct. Under these conditions the assumption is about a quasi-steady state wherein it is *hoped* that $\dot{X}_1(t) << \dot{P}(t)$. Figures 2.2[2] and 2.3[3] illustrate that this hope is not always fulfilled. In Fig. 2.2 appears the relationship

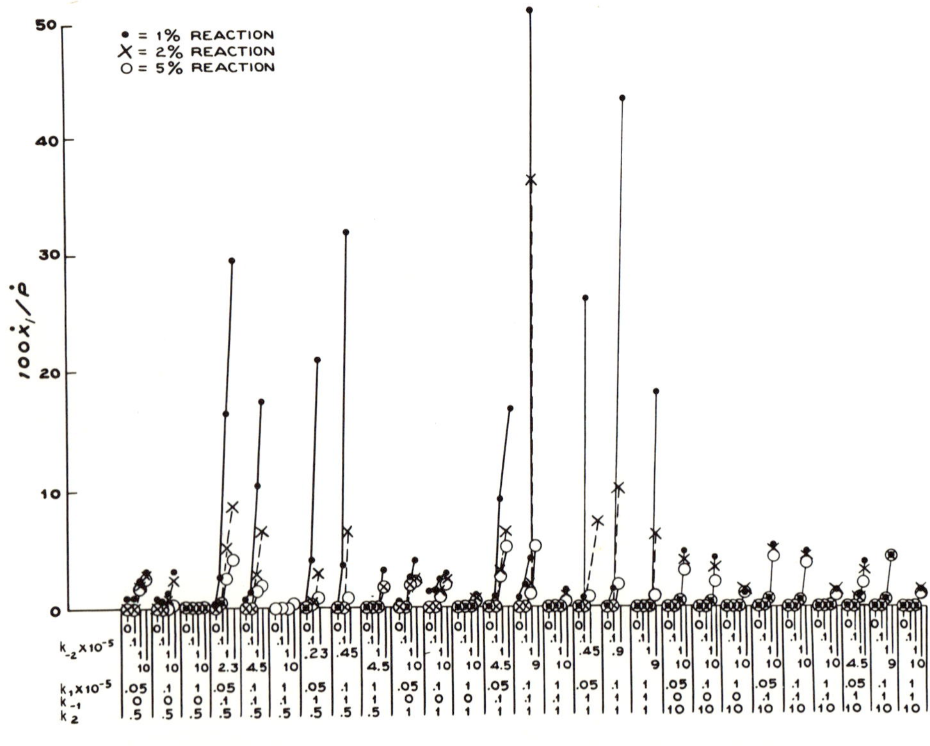

FIG. 2.2. Relationship between the quasi-steady-state error and the values of the individual rate constants in Eqs. (2.1). S(0) = 100 E(0) = 0.1 mM.

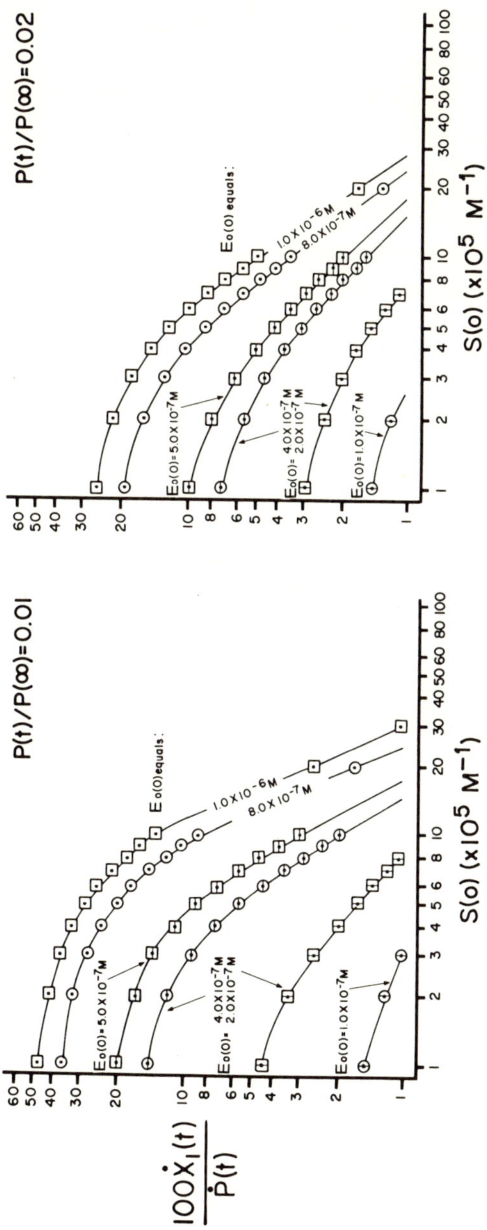

FIG. 2.3. Relationship between the quasi-steady-state error and the initial conditions for Eqs. (2.1). $k_1 = 20/mM \cdot \tau$, $k_{-1} = 0.1/\tau$, $k_2 = 1.9/\tau$, and $k_{-2} = 380/mM \cdot \tau$ where τ is the time unit for the rate constants.

between $|\dot{X}_1(t)/\dot{P}(t)|$ and the four rate constants in Eqs. (2.1). This figure illustrates that when k_1/k_{-2} is in the range 0.01 - 0.1, $\dot{X}_1(t)$ can be 25 - 50% of $\dot{P}(t)$. Presumably, $\dot{X}_1(t)/\dot{P}(t)$ could be even higher for smaller k_1/k_2. In Fig. 2.3 appears the relationship between $|\dot{X}_1(t)/\dot{P}(t)|$, the total enzyme concentration, and the initial substrate concentration for the reactions in Eqs. (2.1). This figure illustrates that when E(0)/S(0) is in the range 0.01 - 0.1, $\dot{X}_1(t)$ can be 25 - 50% of $\dot{P}(t)$. $\dot{X}_1(t)/\dot{P}(t)$ could be even higher for larger E(0)/S(0).

On the other hand, Figs. 2.1, 2.2, and 2.3 also illustrate that *under certain conditions*, $|\dot{X}_1(t)|$ can be considerably smaller than $\dot{P}(t)$ even when the overall reaction is only 1% complete. Under these conditions a quasi-steady-state of $X_1(t)$ arises early during the overall reaction. The validity of the concept of an initial rate depends on t* being sufficiently small, and the validity of the concept of a quasi-steady state depends on the relative values of E(0), S(0), k_1, and k_{-2}.

2.2 THE ANALYTICAL RELATIONSHIP BETWEEN QUASI-STEADY-STATE VALIDITY, RATE CONSTANTS AND INITIAL CONDITIONS

Substitution of $E(0) = E(t) + X_1(t)$ into Eq. (2.2) yields

$$\frac{\dot{X}_1(t)}{k_1\ S(t) + k_{-2}\ P(t) + k_{-1} + k_2} + X_1(t) = \frac{k_1\ S(t) + k_{-2}\ P(t)}{k_1\ S(t) + k_{-2}\ P(t) + k_{-1} + k_2} E(0) \tag{2.8}$$

The quasi-steady-state estimates for $X_1(ss)$ and $\dot{X}_1(ss)$ are

$$X_1(ss) = \frac{k_1\ S(ss) + k_{-2}\ P(ss)}{k_1\ S(ss) + k_{-2}\ P(ss) + k_{-1} + k_2} E(0) \tag{2.9}$$

$$\dot{X}_1(ss) = \frac{(k_1 - k_{-2})(k_{-1} + k_2)\ \dot{S}(ss)\ E(0)}{[k_1\ S(ss) + k_{-2}\ P(ss) + k_{-1} + k_2]^2} \tag{2.10}$$

If we substitute these quasi-steady-state estimates into Eq. (2.8), we obtain

$$1 + \varepsilon \simeq 1 \tag{2.11}$$

where ε is the error introduced by the use of the quasi-steady-state approximation

$$\varepsilon = \frac{(k_1 - k_{-2})(k_{-1} + k_2)\ \dot{S}(ss)\ E(0)}{[k_1\ S(ss) + k_{-2}\ P(ss)][k_1\ S(ss) + k_{-2}\ P(ss) + k_{-1} + k_2]^2} \tag{2.12}$$

The quasi-steady-state estimate for $\dot{S}(ss)$ is

$$-\dot{S}(ss) = \frac{\left[\dfrac{k_2\ S(ss)}{K_f} - \dfrac{k_{-1}\ P(ss)}{K_r}\right] E(0)}{1 + \dfrac{S(ss)}{K_f} + \dfrac{P(ss)}{K_r}} \tag{2.13}$$

Substitution into Eq. (2.12) yields

$$\varepsilon = \frac{\dfrac{k_2}{k_2 + k_{-1}}\dfrac{S(ss)}{K_f} - \dfrac{k_{-1}}{k_2 + k_{-1}}\dfrac{P(ss)}{K_r}}{\left[1 + \dfrac{S(ss)}{K_f} + \dfrac{P(ss)}{K_r}\right]^3}\ \frac{(k_{-2} - k_1)}{k_1\ S(ss) + k_{-2}\ P(ss)}\ E(0) \tag{2.14}$$

When $k_1 = k_{-2}$, $\varepsilon = 0$, and the quasi-steady-state approximation has maximum validity. For this special case, Eq. (2.2) can be integrated analytically:

$$X_1(t) = \frac{2k_1\ E(0)\ S(0)\ (1 - \exp\{-[b^2 - 4k_1{}^2\ E(0)\ S(0)]^{1/2}\ t\})}{[b^2 - 4k_1{}^2\ E(0)\ S(0)]^{1/2} + b}$$
$$+ [\{b^2 - 4k_1{}^2\ E(0)\ S(0)]^{1/2} - b\}$$
$$\cdot \exp\{-[b^2 - 4k_1{}^2\ E(0)\ S(0)]^{1/2}\ t\} \tag{2.15}$$

where $b = k_1 E(0) + k_1 S(0) + k_{-1} + k_2$. If $E(0) << S(0)$, Eq. (2.15) becomes

$$X_1(t) = \frac{k_1 E(0) S(0) (1 - \exp\{-[k_1 S(0) + k_{-1} + k_2] t\})}{k_1 S(0) + k_{-1} + k_2} \tag{2.16}$$

Differentiation of Eq. (2.16) yields

$$\dot{X}_1(t) = k_1 E(0) S(0) \exp\{-[k_1 S(0) + k_{-1} + k_2] t\} \tag{2.17}$$

Equations (2.15) and (2.17) illustrate that $\dot{X}_1(t) > 0$ for all $0 \leq t \leq \infty$. The initial value for $\dot{X}_1(0)$ is $k_1 E(0) S(0)$, but according to Eq. (2.17) $\dot{X}_1(t)$ falls to a very small percent of $\dot{X}_1(0)$ in an extremely short time. Furthermore, according to Eq. (2.16) $X_1(t)$ grows very quickly to within 99% of $X_1(\infty)$ during this short interval. For example, if $k_1 = k_{-2} = 10^8/M\cdot\text{sec}$, $k_{-1} = k_2 = 10^4/\text{sec}$, $S(0) = 10^{-4}\ M$, and $E(0) = 10^{-7}\ M$, then $\dot{X}_1(t) = 10^{-3} \cdot e^{-30000t}$. When $t = 0$, $\dot{X}_1(0) = 1$ mM/sec and $X_1(0) = 0$, but 1 msec later, $\dot{X}_1(0.001) = e^{-30}$ mM/sec and $X_1(0.001) = (1/3)(10^{-4}) \cdot (1 - e^{-30})$ mM. Since $X_1(\infty)$ is $(1/3)(10^{-4})$ mM, $X_1(0.001)$ is imperceptibly smaller than $X_1(\infty)$. Obviously when $k_1 = k_{-2}$, the quasi-steady-state assumption is a good approximation.

When $k_1 > k_{-2}$, $\varepsilon < 0$, and according to Eq. (2.14), the maximum value for $-\varepsilon$ occurs when k_{-2} is zero:

$$-\varepsilon < \frac{\dfrac{k_2}{k_2 + k_{-1}} \dfrac{S(ss)}{K_f}}{\left[1 + \dfrac{S(ss)}{K_f}\right]^3} \frac{E(0)}{S(ss)} \tag{2.18}$$

Since $0 < k_2/(k_2 + k_{-1}) < 1$,

$$-\varepsilon < \frac{\dfrac{S(ss)}{K_f}}{\left[1 + \dfrac{S(ss)}{K_f}\right]^3} \frac{E(0)}{S(ss)} \tag{2.19}$$

For any given E(0)/S(ss) the value of $S(ss)/K_f$ giving the maximum possible value for $-\varepsilon$ can be calculated from

$$\frac{\partial(-\varepsilon)}{\partial\frac{S(ss)}{K_f}} = \frac{1 - 2\frac{S(ss)}{K_f}}{\left[1 + \frac{S(ss)}{K_f}\right]^4} \frac{E(0)}{S(ss)} \tag{2.20}$$

by setting Eq. (2.20) equal to zero and solving for $S(ss)/K_f$:

$$\frac{S(ss)}{K_f} = \frac{1}{2} \tag{2.21}$$

Thus, when $k_1 \geq k_{-2}$, the maximum value possible for $-\varepsilon$ is

$$\varepsilon_{max} = \frac{4}{27}\frac{E(0)}{S(ss)} \qquad k_1 \geq k_{-2} \tag{2.22}$$

According to Eq. (2.22) the maximum error possible due to the quasi-steady-state approximation for Eqs. (2.1) when $k_1 \geq k_{-2}$ is 4/27 of the E(0)/S(ss) ratio[4]. A maximum error of 1% is expected when E(0)/S(ss) is about 1/15; a proportionately smaller error is anticipated for smaller ratios of enzyme to substrate.

When $k_1 < k_{-2}$, then $\varepsilon > 0$, and the maximum value possible for ε is

$$\varepsilon_{max} = \frac{4}{27}\frac{E(0)}{S(ss)}\frac{k_{-2}}{k_1} \qquad [k_1 < k_{-2}] \tag{2.23}$$

According to Eq. (2.23) the maximum error due to applying the quasi-steady-state approximation to the mechanism described by Eqs. (2.1) when $k_1 < k_{-2}$ is 4/27 of the E(0)/S(ss) ratio *times* the ratio of k_{-2}/k_1.[4] If $k_{-2} = 5k_1$, a maximum error of 1% is expected when E(0)/S(ss) is about 1/75. However, if $k_{-2} = 1000k_1$, E(0)/S(ss) would have to be in the range of 15,000 to be sure that an error of 1% was not introduced with the quasi-steady-state approximation. Note that if $k_{-2} = 1000k_1$ and $k_2 = 1000k_{-1}$, the equilibrium constant for the overall reaction would be unity, and one would not know *a priori* that k_{-2} was so much larger than k_1.

CHAPTER 3
QUASI-STEADY-STATE ENZYME MODELS

3.1 THE RELATIONSHIP BETWEEN REAL AND QUASI-STEADY-STATE MODELS

An enzyme mechanism is a chemical model for which there is an exact mathematical description:

$$M(\dot{x}, x, t) = 0 \tag{3.1}$$

In Eq. (3.1) x represents the concentrations of the chemical constituents, and M is the function that describes the model *exactly* in terms of the parameters (the rate constants), the initial conditions, the x, and t (time). A function x* that satisfies Eq. (3.1) is called an exact solution for the model.

It is not uncommon to find that the mathematical representation of a model is facilitated if one makes an approximation. Such approximations introduce errors into the original model in the sense that the approximate mathematical representation

$$M(\dot{z}, z, t) < \varepsilon \tag{3.2}$$

is no longer an exact description of the actual model. Since (3.2) is an inequality, there is not a single solution, but there is a range of z which fulfills the conditions imposed by the magnitude of ε. Since ε should be small and positive, the representation is approximately correct, and we refer to it as being ε valid.

The most common approximation made for enzyme models is the

quasi-steady-state assumption. This approximation is considered valid if ε in inequality (3.2) is sufficiently small (i.e., if more than a maximum acceptable error is not introduced into the model). In the first part of this chapter it is assumed that the mathematical description of the real enzyme mechanism is nearly identical to the approximate description of the quasi-steady-state mechanism. In the last part of this chapter a method of allocating error to model approximation as well as experimental inaccuracy is developed. Application of this method to the general quasi-steady-state model for enzymes leads to a basis for assigning errors to the enzyme model, as well as to experimental determinations.

3.2 EQUATIONS RELATING QUASI-STEADY-STATE RATES AND INITIAL SUBSTRATE CONCENTRATIONS

An important general method for deriving quasi-steady-state rate equations for enzyme mechanisms[5] utilizes Cramer's rule in a relatively uncomplicated way. For an enzyme mechanism involving $n + 1$ different enzyme forms (X_i; $i = 0, 1, \ldots, n$) including $X_0 \equiv E$, the concentration of each X_i relative to the total enzyme concentration $X_i/E(0)$ is a quotient of two summation terms. Each term is the product of n different unimolecular rate constants or bimolecular rate constants times a reactant concentration. Each term in the numerator of $X_m/E(0)$ involves the rate constants (and the reactants for bimolecular constants) associated with reaction steps that individually or in sequence lead to X_m. The n rate constants in each term are associated with the n reaction steps in which each of the n different X_i $[i \neq m]$ is a reactant. All of the possible combinations of n rate constants that conform to this statement are present in the numerator of $X_m/E(0)$. The denominator is the sum of all the various numerators for $m = 0, 1, \ldots, n$.

The following two examples should help to illustrate the advantages of this schematic method of deriving quasi-steady-state rate equations. The mechanism involving two enzyme intermediates (n = 2) is

$$X_0 + S \underset{k_{-1}}{\overset{k_1}{\rightleftarrows}} X_1 \underset{k_{-2}}{\overset{k_2}{\rightleftarrows}} X_2 \underset{k_{-3}}{\overset{k_3}{\rightleftarrows}} X_0 + P \tag{3.3}$$

The numerator and denominator terms can be obtained by inspecting a listing (Table 3.1) of the direct and sequential steps that lead to each X_m. The relative concentration of each enzyme-containing species is proportional to the sum of three terms:

$$\frac{X_0}{E(0)} \propto k_{-1}k_3 + k_{-1}k_{-2} + k_2k_3 \tag{3.4}$$

$$\frac{X_1}{E(0)} \propto k_1Sk_{-2} + k_1Sk_3 + k_{-2}k_{-3}P \tag{3.5}$$

$$\frac{X_2}{E(0)} \propto k_2k_{-3}P + k_1Sk_2 + k_{-1}k_{-3}P \tag{3.6}$$

The proportionality constant for Eqs. (3.4)-(3.6) is $1/(X_0 + X_1 + X_2)$. From the quasi-steady-state rate of product formation

TABLE 3.1. Direct and sequential steps leading to X_0, X_1, and X_2 in Eq. (3.3).

m	Direct steps	Sequential steps	
0	$\xleftarrow{k_{-1}}$ $\xrightarrow{k_3}$	$\xleftarrow{k_{-1}}$ $\xleftarrow{k_{-2}}$	$\xrightarrow{k_2}$ $\xrightarrow{k_3}$
1	$\xrightarrow{k_1S}$ $\xleftarrow{k_{-2}}$	$\xrightarrow{k_1S}$ $\xrightarrow{k_3}$	$\xleftarrow{k_{-2}}$ $\xleftarrow{k_{-3}P}$
2	$\xrightarrow{k_2}$ $\xleftarrow{k_{-3}P}$	$\xrightarrow{k_1S}$ $\xrightarrow{k_2}$	$\xleftarrow{k_{-1}}$ $\xleftarrow{k_{-3}P}$

$$\dot{P} = \frac{k_3X_2 - k_{-3}X_0P}{X_0 + X_1 + X_2} E(0) \tag{3.7}$$

we obtain the quasi-steady-state rate equation

$$\dot{P} = \frac{k_1k_2k_3S - k_{-1}k_{-2}k_{-3}P}{[k_{-1}k_3 + k_{-1}k_{-2} + k_2k_3 + k_1(k_{-2} + k_3 + k_2)S + k_{-3}(k_{-2} + k_{-1} + k_2)P]} E(0) \tag{3.8}$$

The mechanism involving three enzyme intermediates (n = 3) is

$$S + X_0 \underset{k_{-1}}{\overset{k_1}{\rightleftharpoons}} X_1 \underset{k_{-2}}{\overset{k_2}{\rightleftharpoons}} X_2 \underset{k_{-3}}{\overset{k_3}{\rightleftharpoons}} X_3 \underset{k_{-4}}{\overset{k_4}{\rightleftharpoons}} X_0 + P \tag{3.9}$$

The direct and sequential steps for each of the X_m in Eqs. (3.9) are listed in Table 3.2. Inspection of this table leads to the relative concentrations for each enzyme-containing compound:

$$\frac{X_0}{E(0)} \propto k_{-1}k_{-2}k_{-3} + k_{-1}k_{-2}k_4 + k_{-1}k_3k_4 + k_2k_3k_4 \tag{3.10}$$

$$\frac{X_1}{E(0)} \propto k_1Sk_{-2}k_{-3} + k_1Sk_{-2}k_4 + k_1Sk_3k_4 + k_{-2}k_{-3}k_{-4}P \tag{3.11}$$

$$\frac{X_2}{E(0)} \propto k_1Sk_2k_{-3} + k_1Sk_2k_4 + k_{-1}k_{-3}k_{-4}P + k_2k_{-3}k_{-4}P \tag{3.12}$$

$$\frac{X_3}{E(0)} \propto k_1Sk_2k_3 + k_{-1}k_{-2}k_{-4}P + k_{-1}k_3k_{-4}P + k_2k_3k_{-4}P \tag{3.13}$$

The proportionality constant for Eqs. (3.10)-(3.13) is $1/(X_0 + X_1 + X_2 + X_3)$. For this case the quasi-steady-state rate of product formation is

$$\dot{P} = \frac{k_4X_3 - k_{-4}X_0P}{X_0 + X_1 + X_2 + X_3} E(0) \tag{3.14}$$

The quasi-steady-state rate equation for Eqs. (3.9) is

TABLE 3.2. Direct and sequential steps leading to X_0, X_1, X_2, and X_3 in Eq. (3.9).

m	Direct steps		Sequential steps	
0	$\xleftarrow[k_{-1}]{}$ $\xrightarrow{k_3}$ $\xrightarrow{k_4}$	$\xleftarrow[k_{-1}]{}$ $\xleftarrow[k_{-2}]{}$ $\xrightarrow{k_4}$	$\xrightarrow{k_2}$ $\xrightarrow{k_3}$ $\xrightarrow{k_4}$	$\xleftarrow[k_{-1}]{}$ $\xleftarrow[k_{-2}]{}$ $\xleftarrow[k_{-3}]{}$
1	$\xrightarrow{k_1S}$ $\xleftarrow[k_{-2}]{}$ $\xleftarrow[k_{-3}]{}$	$\xrightarrow{k_1S}$ $\xleftarrow[k_{-2}]{}$ $\xrightarrow{k_4}$	$\xrightarrow{k_1S}$ $\xrightarrow{k_3}$ $\xrightarrow{k_4}$	$\xleftarrow[k_{-2}]{}$ $\xleftarrow[k_{-3}]{}$ $\xleftarrow[k_{-4}P]{}$
2	$\xrightarrow{k_2}$ $\xleftarrow[k_{-3}]{}$ $\xleftarrow[k_{-4}P]{}$	$\xrightarrow{k_1S}$ $\xrightarrow{k_2}$ $\xleftarrow[k_{-3}]{}$	$\xrightarrow{k_1S}$ $\xrightarrow{k_2}$ $\xrightarrow{k_4}$	$\xleftarrow[k_{-1}]{}$ $\xleftarrow[k_{-3}]{}$ $\xleftarrow[k_{-4}P]{}$
3	$\xleftarrow[k_{-1}]{}$ $\xrightarrow{k_3}$ $\xleftarrow[k_{-4}P]{}$	$\xrightarrow{k_2}$ $\xrightarrow{k_3}$ $\xleftarrow[k_{-4}P]{}$	$\xrightarrow{k_1S}$ $\xrightarrow{k_2}$ $\xrightarrow{k_3}$	$\xleftarrow[k_{-1}]{}$ $\xleftarrow[k_{-2}]{}$ $\xleftarrow[k_{-4}P]{}$

$$\dot{P} = \frac{(k_1k_2k_3k_4S - k_{-1}k_{-2}k_{-3}k_{-4}P)\,E(0)}{\begin{aligned}[k_{-1}k_3k_4 + k_{-1}k_{-2}k_4 + k_2k_3k_4 + k_{-1}k_{-2}k_{-3}\\ + k_1(k_{-2}k_{-3} + k_{-2}k_4 + k_3k_4 + k_2k_{-3} + k_2k_4 + k_2k_3)S\\ + k_{-4}(k_{-2}k_{-3} + k_{-1}k_{-3} + k_{-1}k_3 + k_2k_{-3} + k_{-1}k_{-2} + k_2k_3)P]\end{aligned}} \tag{3.15}$$

Although Eqs. (3.3) and (3.9) involve only one substrate and product, the corresponding quasi-steady-state rate equations can be used for linear sequences of multisubstrate, multiproduct enzyme reactions involving three [Eq. (3.8)] or four [Eq. (3.15)] steps. In each mechanism there could have been up to n + 1 substrates (S_i; i = 0, 1, ..., n) and/or products (P_i; i = 0, 1, ..., n). Each of these substrates would be capable of binding with only one form of the enzyme, and each product would be formed in only one steps. If by convention we designate that each substrate binds the form of the enzyme with the same subscript and each product is formed in the step when the enzyme-compound with the same subscript is generated, we must interchange $S_0 \equiv S$ and $P_0 \equiv P$, and set $k_{i+1}S_i \equiv k_{i+1}$ [i = 1, ..., n] and $k_{-i}P_i \equiv k_{-i}$ [i = 1, ..., n] for any step involving a substrate or product. After these substitutions have been made, the quasi-steady-state rate equations describe the relationship between this rate and the substrate and product concentrations for ordered mechanisms involving three [Eq. (3.8)] or four [Eq. (3.15)] catalyzed steps. For the general enzyme mechanism

$$\begin{aligned} X_0 + S_0 &\underset{k_{-1}}{\overset{k_1}{\rightleftharpoons}} X_1 + P_1\\ X_1 + S_1 &\underset{k_{-2}}{\overset{k_2}{\rightleftharpoons}} X_2 + P_2\\ &\vdots\\ X_n + S_n &\underset{k_{-(n+1)}}{\overset{k_{n+1}}{\rightleftharpoons}} X_0 + P_0 \end{aligned} \tag{3.16}$$

the quasi-steasy-state rate is

$$\dot{P} = \frac{\dfrac{{}^{m}V_f}{{}^{m}K_f} S_m - \dfrac{{}^{m}V_r}{{}^{m}K_r} P_m}{1 + \dfrac{S_m}{{}^{m}K_f} + \dfrac{P_m}{{}^{m}K_r}} \tag{3.17}$$

where ${}^{m}V_f$, the "maximum forward velocity with respect to S_m," ${}^{m}V_r$, the "maximum reverse velocity with respect to P_m," ${}^{m}K_f$, the "forward Michaelis constant with respect to S_m," and ${}^{m}K_r$, the "reverse Michaelis constant with respect to P_m," are constants that may depend on any of the other S_i or P_i $[i \neq m]$.

For an enzyme mechanism with p reversible steps interconverting the n + 1 different enzyme intermediates, the total number of the p steps taken n at a time is

$$\text{TOTAL} = \frac{p!}{n!(p - n)!} \tag{3.18}$$

Equation (3.18) is useful because it provides a check on the number of numerator terms. For example, in Eq. (3.3) $n = 2$ and $p = 3$, so the total number of numerator terms in Eqs. (3.4)-(3.6) is $3!/2!$ or 3; in Eqs. (3.9) $n = 3$ and $p = 4$, so the total number of numerator terms in Eqs. (3.10)-(3.13) is $4!/3!$ or 4.

Enzyme mechanisms with more than one cycle contain redundancies [see, for example, Eqs. (1.29) and (1.30)]. In such mechanisms there are combinations of rate constants that should not be included in the quasi-steady-state rate equation because they do not conform to the King-Altman requirements[5] described in the first paragraph of this section. For example, in the 2S, 1P-random mechanism,

$$X_0 + S_0 \underset{k_{-1}}{\overset{k_1}{\rightleftharpoons}} X_1$$

$$X_0 + S_1 \underset{k_4}{\overset{k_{-4}}{\rightleftharpoons}} X_3$$

$$X_1 + S_1 \underset{k_{-2}}{\overset{k_2}{\rightleftharpoons}} X_2$$

$$X_3 + S_0 \underset{k_3}{\overset{k_{-3}}{\rightleftharpoons}} X_2$$

$$X_2 \underset{k_{-5}}{\overset{k_5}{\rightleftharpoons}} X_0 + P \tag{3.19}$$

n = 3 and p = 5, so there are 5!/3!2! or ten different pathways involving three of the five steps in Eqs. (3.19). These ten patterns are illustrated in Table 3.3. Because Eqs. (3.19) contain an "extra" cycle, two of these pathways do not conform to the conditions for inclusion. The last two patterns in Table 3.3 should not be included because they do not include *four* of the enzyme forms. The eight remaining terms in the numerator for each

TABLE 3.3. Patterns of pathways involving three of the five steps in Eq. (3.19)

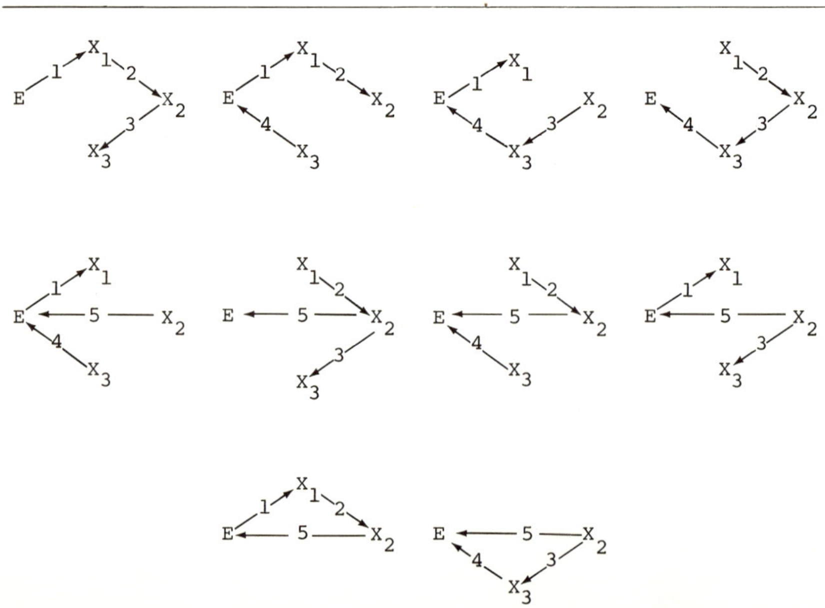

X_m can be written down from inspection of the first eight patterns in Table 3.3:

$$\frac{X_0}{E(0)} \propto k_{-1}k_{-2}k_{-3}S_0 + k_{-1}k_{-2}k_4 + k_{-1}k_3k_4 + k_2S_1k_3k_4 + k_{-1}k_4k_5 + k_2S_1k_4k_5 + k_2S_1k_{-3}S_0k_5 + k_{-1}k_{-3}S_0k_5 \quad (3.20)$$

$$\frac{X_1}{E(0)} \propto k_1S_0k_{-1}k_{-3}S_0 + k_1S_0k_{-2}k_4 + k_1S_0k_3k_4 + k_{-2}k_{-3}S_0k_{-4}S_1 + k_1S_0k_4k_5 + k_{-2}k_4k_{-5}P + k_{-2}k_{-3}S_0k_{-5}P + k_1S_0k_{-3}S_0k_5 \quad (3.21)$$

$$\frac{X_2}{E(0)} \propto k_1S_0k_2S_1k_{-3}S_0 + k_1S_0k_2S_1k_4 + k_{-1}k_{-3}S_0k_{-4}S_1 + k_2S_1k_{-3}S_0k_{-4}S_1 + k_{-1}k_4k_{-5}P + k_2S_1k_4k_{-5}P + k_2S_1k_{-3}S_0k_{-5}P + k_{-1}k_{-3}S_0k_{-5}P \quad (3.22)$$

$$\frac{X_3}{E(0)} \propto k_1S_0k_2S_1k_3 + k_{-1}k_{-2}k_{-4}S_1 + k_{-1}k_3k_{-4}S_1 + k_2S_1k_3k_{-4}S_1 + k_{-1}k_{-4}S_1k_5 + k_2S_1k_{-4}S_1k_5 + k_2S_1k_3k_{-5}P + k_{-1}k_3k_{-5}P \quad (3.23)$$

The quasi-steady-state rate

$$\dot{P} = k_5X_2 - k_{-5}X_0P \quad (3.24)$$

is

$$\dot{P} = \frac{a_5S_0S_1 + a_6P + a_7S_0P + a_8S_1P + a_{10}S_0^2S_1 + a_{12}S_0S_1^2}{(b_0 + b_1S_0 + b_2S_1 + b_3S_0^2 + b_4S_1^2 + b_5S_0S_1 + b_6P + b_7S_0P + b_8S_1P + b_{10}S_0^2S_1 + b_{12}S_0S_1^2 + b_{13}S_0S_1P)} E(0) \quad (3.25)$$

where

$$a_5 = k_5(k_1k_2k_4 + k_{-1}k_{-3}k_{-4})$$

$$a_6 = -k_{-1}k_4k_{-5}(k_{-2} + k_3)$$

$$a_7 = -k_{-1}k_{-2}k_{-3}k_{-5}$$

$$a_8 = -k_2k_3k_4k_{-5}$$

$$a_{10} = k_1k_2k_{-3}k_5$$

$$a_{12} = k_2k_{-3}k_{-4}k_5$$

$$b_0 = k_{-1}k_4(k_{-2} + k_3 + k_5)$$

$$b_1 = k_1k_{-2}k_4 + k_1k_3k_4 + k_1k_4k_5 + k_{-1}k_{-2}k_{-3} + k_{-1}k_{-3}k_5$$

$$b_2 = k_{-1}k_{-2}k_{-4} + k_{-1}k_3k_{-4} + k_{-1}k_{-4}k_5 + k_2k_3k_4 + k_2k_4k_5$$

$$b_3 = k_1k_{-3}(k_{-1} + k_5)$$

$$b_4 = k_2k_{-4}(k_3 + k_5)$$

$$b_5 = k_1k_2k_3 + k_1k_2k_4 + k_{-1}k_{-3}k_{-4} + k_2k_{-3}k_5 + k_{-2}k_{-3}k_{-4}$$

$$b_6 = k_{-5}(k_{-1}k_3 + k_{-1}k_4 + k_{-2}k_4)$$

$$b_7 = k_{-3}k_{-5}(k_{-1} + k_{-2})$$

$$b_8 = k_2k_{-5}(k_3 + k_4)$$

$$b_{10} = k_1k_2k_{-3}$$

$$b_{12} = k_2k_{-3}k_{-4}$$

$$b_{13} = k_2k_{-3}k_{-5}$$

Equation (3.25) can be written in the form

$$\dot{P} = \frac{A_0 + A_1x + A_2x^2}{B_0 + B_1x + B_2x^2} \tag{3.26}$$

for $x = S_0$ or $x = S_1$. In the former case, the coefficients are

$$A_0 = a_6P + a_8S_1P$$

$$A_1 = a_5S_1 + a_7P + a_{12}S_1^2$$

$$A_2 = a_{10}S_1$$

$$B_0 = b_0 + b_2S_1 + b_4S_1^2 + b_6P + b_8S_1P$$

$$B_1 = b_1 + b_5S_1 + b_7P + b_{12}S_1^2 + b_{13}S_1P$$

$$B_2 = b_3 + b_{10}S_1$$

When x is S_1, the coefficients are

$$A_0 = a_6P + a_7S_0P$$

$$A_1 = a_5S_0 + a_8P + a_{10}S_0^2$$

$$A_2 = a_{12}S_0$$

$$B_0 = b_0 + b_1S_0 + b_3S_0^2 + b_6P + b_7S_0P$$

$$B_1 = b_2 + b_5S_0 + b_8P + b_{10}S_0^2 + b_{13}S_0P$$

$$B_2 = b_4 + b_{12}S_0$$

It is evident from this relatively simple enzyme mechanism that pencil-and-paper derivations of quasi-steady-state rate equations can be time consuming. However, since the procedure for obtaining them is straightforward, it is easily programmed for computer-based derivations. The implementation of the King-Altman method[4] into a digital computer program[6,7] was a significant development in quasi-steady-state enzyme kinetics because it made all subsequent manual derivations unnecessary. It is certainly no longer "Science" to spend several hours deriving algebraic expressions that a computer can generate *correctly* much more quickly.

The algorithm to determine quasi-steady-state rate equations is presented in Table 3.4 in the symbolism developed by Iverson[8-10]. A description follows.

Symbolisms and Conventions for Table 3.4

$\longleftarrow$ is a replacement operator. The variable (scalar, vector, matrix) on the left is replaced by the value(s) of the expression on the right.

: is a proposition. If the expression on the left of the colon stands in relation to the expression on the right, then the relation has the value true; otherwise, it has the value false. In this algorithm, the relation operator is used to express alternative branches. For example, at step 48 the relation

$$\xleftarrow{<} k : i \xrightarrow{=}$$

appears. If k < i, the branch to the left (to statement 38) occurs; if k = i, the branch to the right (to statement 51) occurs; if neither of the previous propositions is true, the next statement in sequence, 49, is carried out. Note: If there is no relation operator on a branch arrow, the branch always occurs (i.e., it is an unconditional branch); thus, statement 27 always is executed after statement 29.

TABLE 3.4. Program to determine rate-equation coefficients.

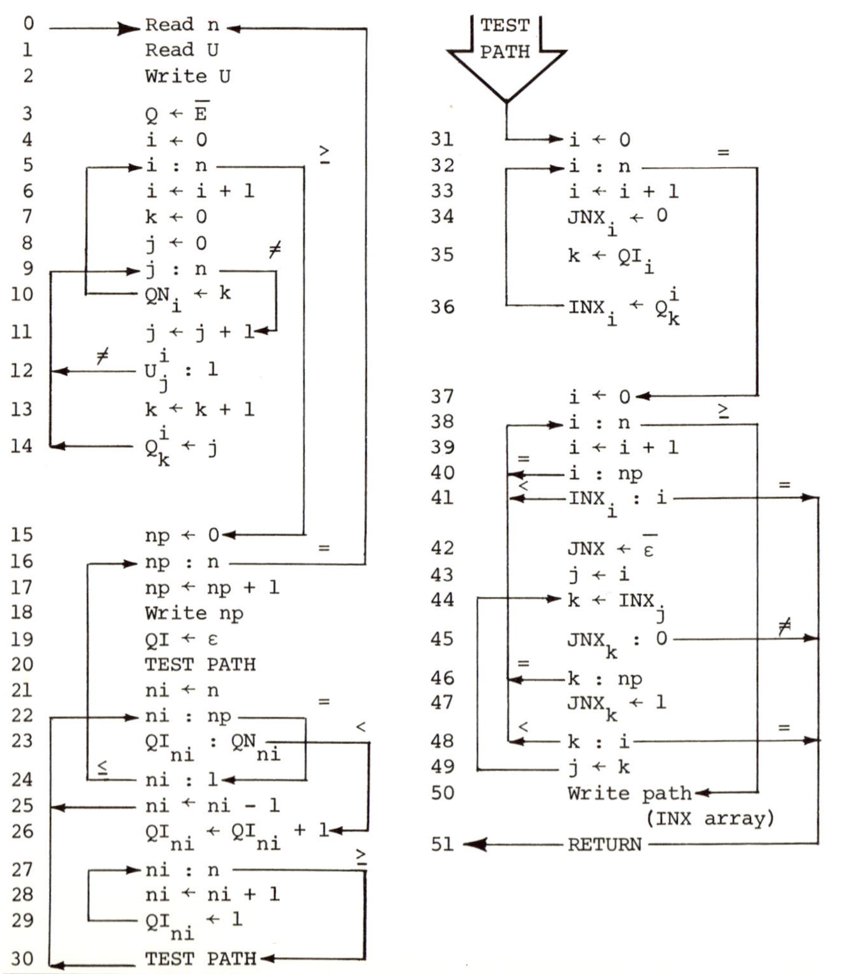

Following is a key to abbreviations used in Table 3.4: The logical operators that the colon may represent are

$=$	equal
$\neq$	not equal
$>$	greater than
$\geq$	greater than or equal to
$<$	less than
$\leq$	less than or equal to

ρ is a dimension operator.

TEST PATH is a subroutine call. Control at statements 20 and 30 transfers to statement 31. When TEST PATH is finished, control returns to the statements after the calling statement.

Legend for Table 3.4

n	number of nodes
np	node number of product node (ranges from 1 to n)
i, j, k, ni	index variables
U; ρ(U) = (n,n)	node connection matrix
Q; ρ(Q) = (n,n)	matrix of nonzero entries in U
QN; ρ(QN) = n	vector of number of nonzero entries in Q
QI; ρ(QI) = n	index vector whose elements QI_j range from 1 to QN_j
INX; ρ(INX) = n	path vector whose elements are permuted combinations of row entries of Q
JNX; ρ(JNX) = n	vector that detects a closed loop within a path
ε; $\rho(\varepsilon)$ = n	vector of all ones
$\bar{\varepsilon}$; $\rho(\bar{\varepsilon})$ = n	vector of all zeros (the complement of ε)
$\bar{E}$; $\rho(\bar{E})$ = (n,n)	matrix of all zeros (the complement of E)

Use of the program to derive quasi-steady-state rates is also straightforward. The computer input consists of a "connection matrix," the number of forward-direction reaction nodes terminating at the free enzyme, and a description of the substrate-binding and product-desorption steps in the mechanism. The connection matrix, U associated with a mechanism consists of elements ${}^{i}u_{j}$, where ${}^{i}u_{j} = 1$ if there is a reaction path from source node i to destination j, but ${}^{i}u_{j} = 0$ otherwise. In the connection matrix the numbers in the rows correspond to the source nodes, and the numbers in the columns to the destination nodes. For example,

$$U = \begin{matrix} 0 & 1 & 1 \\ 1 & 0 & 1 \\ 1 & 1 & 0 \end{matrix}$$

is the connection matrix for Eqs. (3.3). There are three forward-direction reaction nodes terminating at the free enzyme in Eqs. (3.3) and

$$\begin{matrix} X_0 & S) & P) \\ & X_1 & \\ & & X_2 \end{matrix}$$

describes the first step as the substrate-binding step and the last step as the product-desorption step.

$$U = \begin{matrix} 0 & 1 & 0 & 1 \\ 1 & 0 & 1 & 0 \\ 0 & 1 & 0 & 1 \\ 1 & 0 & 1 & 0 \end{matrix}$$

is the connection matrix for Eqs. (3.9), there are four nodes terminating at the free enzyme, and

$$\begin{matrix} X_0 & S) & & P) \\ & X_1 & & \\ & & X_2 & \\ & & & X_3 \end{matrix}$$

describes the mechanism. If there were an additional substrate (R) and product (Q), and if the mechanism were 2S, 2P-ordered-(SSPP), the input description would be

X_0 S) P)

X_1 R)

X_2

Q) X_3

The output from the quasi-steady-state rate-equation computer program consists of the "terminal-node matrix" for each terminal node of the reaction scheme and the coefficients that appear in the numerator of the rate equation. Each row of the terminal-node matrix refers to the group of rate constants that constitute a term in the denominator of the rate equation. The column numbers of this matrix refer to the source node for the step described by the rate constant, and if it is not zero, the numerical value of the individual entry refers to the destination node of the step. If the entry is zero, there is no reaction between the indicated steps for the mechanism under consideration. For example, the terminal-node matrix

+ 0 1 1

+ 0 1 2

+ 0 3 1

is for terminal node 1 (X_0) of the mechanism described by Eqs. (3.3). The plus signs indicate positive coefficients. The first row is $+k_{-1}k_3$ (node 2 to node 1 and node 3 to node 1), the second row is $+k_{-1}k_{-2}$ (node 2 to node 1 and node 3 to node 2), and the third row is $+k_2k_3$ (node 2 to node 3 and node 3 to node 1). Terminal-node matrix 1 therefore yields the three terms

$$+k_{-1}k_3 + k_{-1}k_{-2} + k_2k_3$$

which are the terms whose sum is proportional to X_0 [see Eq. (3.4)].

The terminal-node matrix

S) + 2 0 1

S) + 2 0 2

P) + 3 0 2

is for node 2 (X_1) of the mechanism described by Eqs. (3.3). This matrix yields the three terms

$$+k_1Sk_3 + k_1Sk_{-2} + k_{-2}k_{-3}P$$

which are the terms whose sum is proportional to X_1 [see Eq. (3.5)]. The other terminal-node matrix for the mechanism described by Eqs. (3.3),

S) + 2 3 0
P) + 3 1 0
P) + 3 3 0

is for terminal node 3 (X_2). This matrix yields the terms

$$+k_1Sk_2 + k_{-1}k_{-3}P + k_2k_{-3}P$$

which are the terms whose sum is proportional to X_2 [see Eq. (3.6)].

The denominator of the quasi-steady-state rate equation is the sum of all the terms generated by all *valid* terminal nodes:

$$k_{-1}k_{-2} + k_{-1}k_3 + k_2k_3$$
$$+ k_1(k_2 + k_{-2} + k_3)S + k_{-3}(k_{-1} + k_2 + k_{-2})P$$

The rate-equation coefficients generated for the mechanism described by Eqs. (3.3) are

S) + 2 3 1
P) - 3 1 2

Therefore the numerator for the rate equation is

$$+k_1Sk_2k_3 - k_{-1}k_{-2}k_{-3}P$$

These combinations of rate constants for the numerator and the denominator yield Eq. (3.8), the quasi-steady-state rate equation for the mechanism described by Eqs. (3.3).

The matrix for terminal node 1 (X_0) of the mechanism described by Eq. (3.9) is

+ 0 1 2 1

+ 0 1 2 3

+ 0 1 4 1

+ 0 3 4 1

which corresponds to the four terms in Eq. (3.10). The matrix for terminal node 2 (X_1) is

S) + 2 0 2 1

S) + 2 0 2 3

S) + 2 0 4 1

P) + 4 0 2 3

This matrix corresponds to the four terms in Eq. (3.11). The matrix for terminal node 3 (X_2) is

S) + 2 3 0 1

S) + 2 3 0 3

P) + 4 1 0 3

P) + 4 3 0 3

This corresponds to the terms in Eq. (3.12). The matrix for terminal node 4 (X_3) is

S) + 2 3 4 0

P) + 4 1 2 0

P) + 4 1 4 0

P) + 4 3 4 0

This matrix corresponds to the terms in Eq. (3.13). The rate-equation coefficients generated for the mechanism described by Eqs. (3.9) are

S) + 2 3 4 1

P) - 4 1 2 3

These terms correspond to the numerator in Eq. (3.15), and the sum of the sixteen terms generated in the four terminal-node matrices correspond to the denominator of this equation.

The power of using the computer-based derivation procedure for generating quasi-steady-state rate equations can be illustrated by applying the method to a realistic general enzyme mechanism. The simplest 2S, 2P-random or 2S, 2P-preferred mechanism is

$$X_0 + S_0 \underset{k_{-1}}{\overset{k_1}{\rightleftharpoons}} X_1$$

$$X_1 + S_1 \underset{k_{-2}}{\overset{k_2}{\rightleftharpoons}} X_2$$

$$X_0 + S_1 \underset{k_{4}}{\overset{k_{-4}}{\rightleftharpoons}} X_3$$

$$X_3 + S_0 \underset{k_{3}}{\overset{k_{-3}}{\rightleftharpoons}} X_2$$

$$X_2 \underset{k_{-5}}{\overset{k_5}{\rightleftharpoons}} X_4$$

$$X_4 \underset{k_{-6}}{\overset{k_6}{\rightleftharpoons}} X_5 + P_0$$

$$X_5 \underset{k_{-7}}{\overset{k_7}{\rightleftharpoons}} X_0 + P_1$$

$$X_4 \underset{k_{9}}{\overset{k_{-9}}{\rightleftharpoons}} X_6 + P_1$$

$$X_6 \underset{k_{8}}{\overset{k_{-8}}{\rightleftharpoons}} X_0 + P_0 \qquad (3.27)$$

The connection pathways and terminal node numbers for this mechanism appear in Table 3.5. The connection matrix input for the quasi-steady-state rate equation program is

TABLE 3.5. Connection pathways and terminal node numbers for Eqs. (3.27).

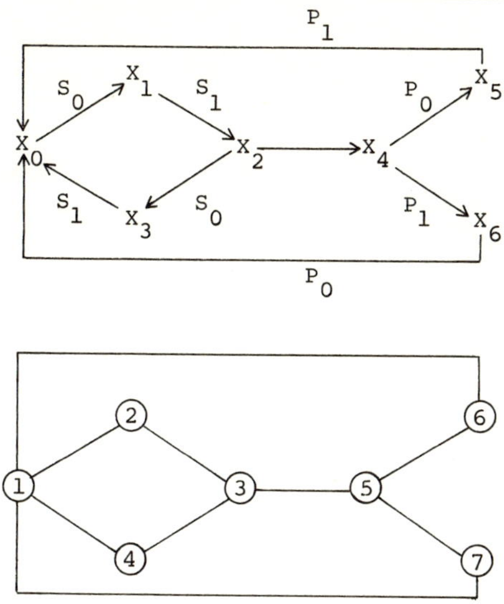

0	1	0	1	0	1	1
1	0	1	0	0	0	0
0	1	0	1	1	0	0
1	0	1	0	1	0	0
0	0	0	1	0	1	1
1	0	0	0	1	0	0
1	0	0	0	1	0	0

and the substrate and product identification input is

X_0	S_0)	S_1)			P_1)	P_0)
	X_1		S_1)			
		X_3	S_0)			
			X_2			
				X_4		
				P_0)	X_5	
				P_1)		X_6

The computer output for terminal nodes 1-7 and for the rate-equation numerator appears in Table 3.6. There are 32 numerator and 336 denominator terms in the quasi-steady-state rate equation:

$$
\frac{\dot{P}}{E(0)} = \frac{\begin{aligned}[t] &[a_6 S_0 S_1 + a_{13} P_0 P_1 + a_{15} S_0^2 S_1 \\ &+ a_{18} S_0 S_1^2 + a_{19} S_0 S_1 P_0 + a_{20} S_0 S_1 P_1 \\ &+ a_{22} S_0 P_0 P_1 + a_{27} S_1 P_0 P_1 + a_{29} P_0^2 P_1 + a_{30} P_0 P_1^2 \\ &+ a_{31} S_0^2 S_1 P_0 + a_{32} S_0^2 S_1 P_1 + a_{34} S_0 S_1^2 P_0 + a_{35} S_0 S_1^2 P_1 \\ &+ a_{39} S_0 P_0^2 P_1 + a_{40} S_0 P_0 P_1^2 + a_{42} S_1 P_0^2 P_1 + a_{43} S_1 P_0 P_1^2] \end{aligned}}{\begin{aligned}[t] &[b_0 + b_1 S_0 + b_2 S_1 + b_3 P_0 + b_4 P_1 + b_5 S_0^2 + b_6 S_0 S_1 \\ &+ b_7 S_0 P_0 + b_8 S_0 P_1 + b_9 S_1^2 + b_{10} S_1 P_0 + b_{11} S_1 P_1 \\ &+ b_{12} P_0^2 + b_{13} P_0 P_1 + b_{14} P_1^2 + b_{15} S_0^2 S_1 + b_{16} S_0^2 P_0 \\ &+ b_{17} S_0^2 P_1 + b_{18} S_0 S_1^2 + b_{19} S_0 S_1 P_0 + b_{20} S_0 S_1 P_1 \\ &+ b_{21} S_0 P_0^2 + b_{22} S_0 P_0 P_1 + b_{23} S_0 P_1^2 + b_{24} S_1^2 P_0 \\ &+ b_{25} S_1^2 P_1 + b_{26} S_1 P_0^2 + b_{27} S_1 P_0 P_1 + b_{28} S_1 P_1^2 \\ &+ b_{29} P_0^2 P_1 + b_{30} P_0 P_1^2 + b_{31} S_0^2 S_1 P_0 + b_{32} S_0^2 S_1 P_1 \\ &+ b_{33} S_0^2 P_0 P_1 + b_{34} S_0 S_1^2 P_0 + b_{35} S_0 S_1^2 P_1 + b_{36} S_0 S_1 P_0^2 \\ &+ b_{37} S_0 S_1 P_0 P_1 + b_{38} S_0 S_1 P_1^2 + b_{39} S_0 P_0^2 P_1 + b_{40} S_0 P_0 P_1^2 \\ &+ b_{41} S_1^2 P_0 P_1 + b_{42} S_1 P_0^2 P_1 + b_{43} S_1 P_0 P_1^2 + b_{44} S_0^2 S_1 P_0 P_1 \\ &+ b_{45} S_0 S_1^2 P_0 P_1 + b_{46} S_0 S_1 P_0^2 P_1 + b_{47} S_0 S_1 P_0 P_1^2] \end{aligned}} \tag{3.28}
$$

The a's and b's in Eq. (3.28) depend only on the rate constants in Eqs. (3.27). The individual definitions for these constants appear in Table 3.7. In this table the matrices for the various rate constants have been grouped according to which concentration(s) they

TABLE 3.6. Terminal nodes and rate equation coefficients for the mechanism described by Eqs. (3.7).

Terminal node is 1								Terminal node is 2							
+	0	1	1	2	4	1	1	S0)+	2	0	1	2	4	1	1
P1)+	0	1	1	2	4	1	5	S0)P1)+	2	0	1	2	4	1	5
P0)+	0	1	1	2	4	5	1	S0)P0)+	2	0	1	2	4	5	1
P0)P1)+	0	1	1	2	4	5	5	S0)P0)P1)+	2	0	1	2	4	5	5
+	0	1	1	2	6	1	1	S0)+	2	0	1	2	6	1	1
P1)+	0	1	1	2	6	1	5	S0)P1)+	2	0	1	2	6	1	5
+	0	1	1	2	7	1	1	S0)+	2	0	1	2	7	1	1
P0)+	0	1	1	2	7	5	1	S0)P0)+	2	0	1	2	7	5	1
+	0	1	1	3	4	1	1	S0)+	2	0	1	3	4	1	1
P1)+	0	1	1	3	4	1	5	S0)P1)+	2	0	1	3	4	1	5
P0)+	0	1	1	3	4	5	1	S0)P0)+	2	0	1	3	4	5	1
P0)P1)+	0	1	1	3	4	5	5	S0)P0)P1)+	2	0	1	3	4	5	5
+	0	1	1	3	6	1	1	S0)+	2	0	1	3	6	1	1
P1)+	0	1	1	3	6	1	5	S0)P1)+	2	0	1	3	6	1	5
+	0	1	1	3	7	1	1	S0)+	2	0	1	3	7	1	1
P0)+	0	1	1	3	7	5	1	S0)P0)+	2	0	1	3	7	5	1
+	0	1	1	5	6	1	1	S0)+	2	0	1	5	6	1	1
P1)+	0	1	1	5	6	1	5	S0)P1)+	2	0	1	5	6	1	5
+	0	1	1	5	7	1	1	S0)+	2	0	1	5	7	1	1
P0)+	0	1	1	5	7	5	1	S0)P0)+	2	0	1	5	7	5	1
S0)+	0	1	4	2	4	1	1	S0)S0)+	2	0	4	2	4	1	1
S0)P1)+	0	1	4	2	4	1	5	S0)S0)P1)+	2	0	4	2	4	1	5
S0)P0)+	0	1	4	2	4	5	1	S0)S0)P0)+	2	0	4	2	4	5	1
S0)P0)P1)+	0	1	4	2	4	5	5	S0)S0)P0)P1)+	2	0	4	2	4	5	5
S0)+	0	1	4	2	6	1	1	S0)S0)+	2	0	4	2	6	1	1
S0)P1)+	0	1	4	2	6	1	5	S0)S0)P1)+	2	0	4	2	6	1	5
S0)+	0	1	4	2	7	1	1	S0)S0)+	2	0	4	2	7	1	1
S0)P0)+	0	1	4	2	7	5	1	S0)S0)P0)+	2	0	4	2	7	5	1
S0)+	0	1	4	5	6	1	1	S0)S0)+	2	0	4	5	6	1	1
S0)P1)+	0	1	4	5	6	1	5	S0)S0)P1)+	2	0	4	5	6	1	5
S0)+	0	1	4	5	7	1	1	S0)S0)+	2	0	4	5	7	1	1
S0)P0)+	0	1	4	5	7	5	1	S0)S0)P0)+	2	0	4	5	7	5	1
S1)+	0	4	1	3	4	1	1	S0)S1)+	3	0	4	2	4	1	1
S1)P1)+	0	4	1	3	4	1	5	S0)S1)P1)+	3	0	4	2	4	1	5
S1)P0)+	0	4	1	3	4	5	1	S0)S1)P0)+	3	0	4	2	4	5	1
S1)P0)P1)+	0	4	1	3	4	5	5	S0)S1)P0)P1)+	3	0	4	2	4	5	5
S1)+	0	4	1	3	6	1	1	S0)S1)+	3	0	4	2	6	1	1
S1)P1)+	0	4	1	3	6	1	5	S0)S1)P1)+	3	0	4	2	6	1	5
S1)+	0	4	1	3	7	1	1	S0)S1)+	3	0	4	2	7	1	1
S1)P0)+	0	4	1	3	7	5	1	S0)S1)P0)+	3	0	4	2	7	5	1
S1)+	0	4	1	5	6	1	1	P0)P1)+	6	0	1	2	4	5	1
S1)P1)+	0	4	1	5	6	1	5	P0)P1)P1)+	6	0	1	2	4	5	5
S1)+	0	4	1	5	7	1	1	S0)P0)P1)+	6	0	4	2	4	5	1
S1)P0)+	0	4	1	5	7	5	1	S0)P0)P1)P1)+	6	0	4	2	4	5	5
S0)S1)+	0	4	4	5	6	1	1	P0)P1)+	7	0	1	2	4	1	5
S0)S1)P1)+	0	4	4	5	6	1	5	P0)P0)P1)+	7	0	1	2	4	5	5
S0)S1)+	0	4	4	5	7	1	1	S0)P0)P1)+	7	0	4	2	4	1	5
S0)S1)P0)+	0	4	4	5	7	5	1	S0)P0)P0)P1)+	7	0	4	2	4	5	5

TABLE 3.6 (continued)

Terminal node is 3							
S0)S1)+	2	4	0	3	4	1	1
S0)S1)P1)+	2	4	0	3	4	1	5
S0)S1)P0)+	2	4	0	3	4	5	1
S0)S1)P0)P1)+	2	4	0	3	4	5	5
S0)S1)+	2	4	0	3	6	1	1
S0)S1)P1)+	2	4	0	3	6	1	5
S0)S1)+	2	4	0	3	7	1	1
S0)S1)P0)+	2	4	0	3	7	5	1
S1)+	3	1	0	2	4	1	1
S1)P1)+	3	1	0	2	4	1	5
S1)P0)+	3	1	0	2	4	5	1
S1)P0)P1)+	3	1	0	2	4	5	5
S1)+	3	1	0	2	6	1	1
S1)P1)+	3	1	0	2	6	1	5
S1)+	3	1	0	2	7	1	1
S1)P1)+	3	1	0	2	7	5	1
S1)+	3	1	0	3	4	1	1
S1)P1)+	3	1	0	3	4	1	5
S1)P0)+	3	1	0	3	4	5	1
S0)P0)P1)+	3	1	0	3	4	5	5
S1)+	3	1	0	3	6	1	1
S1)P1)+	3	1	0	3	6	1	5
S1)+	3	1	0	3	7	1	1
S1)P0)+	3	1	0	3	7	5	1
S1)+	3	1	0	5	6	1	1
S1)P1)+	3	1	0	5	6	1	5
S1)+	3	1	0	5	7	1	1
S1)P0)+	3	1	0	5	7	5	1
S1)S1)+	3	4	0	3	4	1	1
S1)S1)P1)+	3	4	0	3	4	1	5
S1)S1)P0)+	3	4	0	3	4	5	1
S1)S1)P0)P1)+	3	4	0	3	4	5	5
P0)P0)+	3	4	0	3	6	1	1
S1)S1)P0)+	3	4	0	3	6	1	5
S1)S1)+	3	4	0	3	7	1	1
S1)S1)P0)+	3	4	0	3	7	5	1
S1)S1)+	3	4	0	5	6	1	1
S1)S1)P1)+	3	4	0	5	6	1	5
S1)S1)+	3	4	0	5	7	1	1
S1)S1)P0)+	3	4	0	5	7	5	1
P0)P1)+	6	1	0	3	4	5	1
P0)P1)P1)+	6	1	0	3	4	5	5
S1)P0)P1)+	6	4	0	3	4	5	1
S1)P0)P1)P1)+	6	4	0	3	4	5	5
P0)P1)+	7	1	0	3	4	1	5
P0)P0)P1)+	7	1	0	3	4	5	5
S1)P0)P1)+	7	4	0	3	4	1	5
S1)P0)P0)P1)+	7	4	0	3	4	5	5

Terminal node is 4							
S0)S1)+	2	4	1	0	4	1	1
S0)S1)P1)+	2	4	1	0	4	1	5
S0)S1)P0)+	2	4	1	0	4	5	1
S0)S1)P0)P1)+	2	4	1	0	4	5	5
S0)S1)+	2	4	1	0	6	1	1
S0)S1)P1)+	2	4	1	0	6	1	5
S0)S1)+	2	4	1	0	7	1	1
S0)S1)P0)+	2	4	1	0	7	5	1
S0)S0)S1)+	2	4	4	0	4	1	1
S0)S0)S1)P1)+	2	4	4	0	4	1	5
S0)S0)S1)P0)+	2	4	4	0	4	5	1
S0)S0)S1)P0)P1)+	2	4	4	0	4	5	5
S0)S0)S1)+	2	4	4	0	6	1	1
S0)S0)S1)P1)+	2	4	4	0	6	1	5
S0)S0)S1)+	2	4	4	0	7	1	1
S0)S0)S1)P0)+	2	4	4	0	7	5	1
S0)S1)+	3	1	4	0	4	1	1
S0)S1)P1)+	3	1	4	0	4	1	5
S0)S1)P0)+	3	1	4	0	4	5	1
S0)S1)P0)P1)+	3	1	4	0	4	5	5
S0)S1)+	3	1	4	0	6	1	1
S0)S1)P1)+	3	1	4	0	6	1	5
S0)S1)+	3	1	4	0	7	1	1
S0)S1)P0)+	3	1	4	0	7	5	1
S0)S1)S1)+	3	4	4	0	4	1	1
S0)S1)S1)P1)+	3	4	4	0	4	1	5
S0)S1)S1)P0)+	3	4	4	0	4	5	1
S0)S1)S1)P0)P1)+	3	4	4	0	4	5	5
S0)S1)S1)+	3	4	4	0	6	1	1
S0)S1)S1)P1)+	3	4	4	0	6	1	5
S0)S1)S1)+	3	4	4	0	7	1	1
S0)S1)S1)P0)+	3	4	4	0	7	5	1
P0)P1)+	6	1	1	0	4	5	1
P0)P1)P1)+	6	1	1	0	4	5	5
S0)P0)P1)+	6	1	4	0	4	5	1
S0)P0)P1)P1)+	6	1	4	0	4	5	5
S1)P0)P1)+	6	4	1	0	4	5	1
S1)P0)P1)P1)+	6	4	1	0	4	5	5
S0)S1)P0)P1)+	6	4	4	0	4	5	1
S0)S1)P0)P1)P1)+	6	4	4	0	4	5	5
P0)P1)+	7	1	1	0	4	1	5
P0)P0)P1)+	7	1	1	0	4	5	5
S0)P0)P1)+	7	1	4	0	4	1	5
S0)P0)P0)P1)+	7	1	4	0	4	5	5
S1)P0)P1)+	7	4	1	0	4	1	5
S1)P0)P0)P1)+	7	4	1	0	4	5	5
S0)S1)P0)P1)+	7	4	4	0	4	1	5
S0)S1)P0)P0)P1)+	7	4	4	0	4	5	5

TABLE 3.6 (continued)

Terminal node is 5							
S0)S1)+	2	4	1	5	0	1	1
S0)S1)P1)+	2	4	1	5	0	1	5
S0)S1)P0)+	2	4	1	5	0	5	1
S0)S1)P0)P1)+	2	4	1	5	0	5	5
S0)S0)S1)+	2	4	4	5	0	1	1
S0)S0)S1)P1)+	2	4	4	5	0	1	5
S0)S0)S1)P0)+	2	4	4	5	0	5	1
S0)S0)S1)P0)P1)+	2	4	4	5	0	5	5
S0)S1)+	3	1	4	5	0	1	1
S0)S1)P1)+	3	1	4	5	0	1	5
S0)S1)P0)+	3	1	4	5	0	5	1
S0)S1)P0)P1)+	3	1	4	5	0	5	5
S0)S1)S1)+	3	4	4	5	0	1	1
S0)S1)S1)P1)+	3	4	4	5	0	1	5
S0)S1)S1)P0)+	3	4	4	5	0	5	1
S0)S1)S1)P0)P1)+	3	4	4	5	0	5	5
P0)P1)+	6	1	1	2	0	5	1
P0)P1)P1)+	6	1	1	2	0	5	5
P0)P1)+	6	1	1	3	0	5	1
P0)P1)P1)+	6	1	1	3	0	5	5
P0)P1)+	6	1	1	5	0	5	1
P0)P1)P1)+	6	1	1	5	0	5	5
S0)P0)P1)+	6	1	4	2	0	5	1
S0)P0)P1)P1)+	6	1	4	2	0	5	5
S0)P0)P1)+	6	1	4	5	0	5	1
S0)P0)P1)P1)+	6	1	4	5	0	5	5
S1)P0)P1)+	6	4	1	3	0	5	1
S1)P0)P1)P1)+	6	4	1	3	0	5	5
S1)P0)P1)+	6	4	1	5	0	5	1
S1)P0)P1)P1)+	6	4	1	5	0	5	5
S0)S1)P0)P1)+	6	4	4	5	0	5	1
S0)S1)P0)P1)P1)+	6	4	4	5	0	5	5
P0)P1)+	7	1	1	2	0	1	5
P0)P0)P1)+	7	1	1	2	0	5	5
P0)P1)+	7	1	1	3	0	1	5
P0)P0)P1)+	7	1	1	3	0	5	5
P0)P1)+	7	1	1	5	0	1	5
P0)P0)P1)+	7	1	1	5	0	5	5
S0)P0)P1)+	7	1	4	2	0	1	5
S0)P0)P0)P1)+	7	1	4	2	0	5	5
S0)P0)P1)+	7	1	4	5	0	1	5
S0)P0)P0)P1)+	7	1	4	5	0	5	5
S1)P0)P1)+	7	4	1	3	0	1	5
S1)P0)P0)P1)+	7	4	1	3	0	5	5
S1)P0)P1)+	7	4	1	5	0	1	5
S1)P0)P0)P1)+	7	4	1	5	0	5	5
S0)P0)P0)P1)+	7	4	4	5	0	1	5
S0)S1)P0)P0)P1)+	7	4	4	5	0	5	5

Terminal node is 6							
S0)S1)+	2	4	1	5	6	0	1
S0)S1)P1)+	2	4	1	5	6	0	5
S0)S0)S1)+	2	4	4	5	6	0	1
S0)S0)S1)P1)+	2	4	4	5	6	0	5
S0)S1)+	3	1	4	5	6	0	1
S0)S1)P1)+	3	1	4	5	6	0	5
S0)S1)S1)+	3	4	4	5	6	0	1
S0)S1)S1)P1)+	3	4	4	5	6	0	5
P1)+	6	1	1	2	4	0	1
P1)P1)+	6	1	1	2	4	0	5
P1)+	6	1	1	2	6	0	1
P1)P1)+	6	1	1	2	6	0	5
P1)+	6	1	1	2	7	0	1
P1)+	6	1	1	3	4	0	1
P1)P1)+	6	1	1	3	4	0	5
P1)+	6	1	1	3	6	0	1
P1)P1)+	6	1	1	3	6	0	5
P1)+	6	1	1	3	7	0	1
P1)+	6	1	1	5	6	0	1
P1)P1)+	6	1	1	5	6	0	5
P1)+	6	1	1	5	7	0	1
S0)P1)+	6	1	4	2	4	0	1
S0)P1)P1)+	6	1	4	2	4	0	5
S0)P1)+	6	1	4	2	6	0	1
S0)P1)P1)+	6	1	4	2	6	0	5
S0)P1)+	6	1	4	2	7	0	1
S0)P1)+	6	1	4	5	6	0	1
S0)P1)P1)+	6	1	4	5	6	0	5
S0)P1)+	6	1	4	5	7	0	1
S1)P1)+	6	4	1	3	4	0	1
S1)P1)P1)+	6	4	1	3	4	0	5
S1)P1)+	6	4	1	3	6	0	1
S1)P1)P1)+	6	4	1	3	6	0	5
S1)P1)+	6	4	1	3	7	0	1
S1)P1)+	6	4	1	5	6	0	1
S1)P1)P1)+	6	4	1	5	6	0	5
S1)P1)+	6	4	1	5	7	0	1
S0)S1)P1)+	6	4	4	5	6	0	1
S0)S1)P1)P1)+	6	4	4	5	6	0	5
S0)S1)P1)+	6	4	4	5	7	0	1
P0)P1)+	7	1	1	2	6	0	5
P0)P1)+	7	1	1	3	6	0	5
P0)P1)+	7	1	1	5	6	0	5
S0)P0)P1)+	7	1	4	2	6	0	5
S0)P0)P1)+	7	1	4	5	6	0	5
S1)P0)P1)+	7	4	1	3	6	0	5
S1)P0)P1)+	7	4	1	5	6	0	5
S0)S1)P0)P1)+	7	4	4	5	6	0	5

TABLE 3.6 (continued)

Terminal node is 7							
S0)S1)+	2	4	1	5	7	1	0
S0)S1)P0)+	2	4	1	5	7	5	0
S0)S0)S1)+	2	4	4	5	7	1	0
S0)S0)S1)P0)+	2	4	4	5	7	5	0
S0)S1)+	3	1	4	5	7	1	0
S0)S1)P0)+	3	1	4	5	7	5	0
S0)S1)S1)+	3	4	4	5	7	1	0
S0)S1)S1)P0)+	3	4	4	5	7	5	0
P0)P1)+	6	1	1	2	7	5	0
P0)P1)+	6	1	1	3	7	5	0
P0)P1)+	6	1	1	5	7	5	0
S0)P0)P1)+	6	1	4	2	7	5	0
S0)P0)P1)+	6	1	4	5	7	5	0
S1)P0)P1)+	6	4	1	3	7	5	0
S1)P0)P1)+	6	4	1	5	7	5	0
S0)S1)P0)P1)+	6	4	4	5	7	5	0
P0)+	7	1	1	2	4	1	0
P0)P0)+	7	1	1	2	4	5	0
P0)+	7	1	1	2	6	1	0
P0)+	7	1	1	2	7	1	0
P0)P0)+	7	1	1	2	7	5	0
P0)+	7	1	1	3	4	1	0
P0)P0)+	7	1	1	3	4	5	0
P0)+	7	1	1	3	6	1	0
P0)+	7	1	1	3	7	1	0
P0)P0)+	7	1	1	3	7	5	0
P0)+	7	1	1	5	6	1	0
P0)+	7	1	1	5	7	1	0
P0)P0)+	7	1	1	5	7	5	0
S0)P0)+	7	1	4	2	4	1	0
S0)P0)P0)+	7	1	4	2	4	5	0
S0)P0)+	7	1	4	2	6	1	0
S0)P0)+	7	1	4	2	7	1	0
S0)P0)P0)+	7	1	4	2	7	5	0
S0)P0)+	7	1	4	5	6	1	0
S0)P0)+	7	1	4	5	7	1	0
S0)P0)P0)+	7	1	4	5	7	5	0
S1)P0)+	7	4	1	3	4	1	0
S1)P0)P0)+	7	4	1	3	4	5	0
S1)P0)+	7	4	1	3	6	1	0
S1)P0)+	7	4	1	3	7	1	0
S1)P0)P0)+	7	4	1	3	7	5	0
S1)P0)+	7	4	1	5	6	1	0
S1)P0)+	7	4	1	5	7	1	0
S1)P0)P0)+	7	4	1	5	7	5	0
S0)S1)P0)+	7	4	4	5	6	1	0
S0)S1)P0)+	7	4	4	5	7	1	0
S0)S1)P0)P0)+	7	4	4	5	7	5	0

TABLE 3.7. Definitions of the coefficients in Eq. (3.28).

Rate equation numerator

S0)S1)+	2	4	1	5	6	1	1
S0)S1)+	3	1	4	5	6	1	1
S0)S1)+	2	4	1	5	7	1	1
S0)S1)+	3	1	4	5	7	1	1
S0)S0)S1)+	2	4	4	5	6	1	1
S0)S0)S1)+	2	4	4	5	7	1	1
S0)S1)S1)+	3	4	4	5	6	1	1
S0)S1)S1)+	3	4	4	5	7	1	1
S0)S1)P0)+	2	4	1	5	7	5	1
S0)S1)P0)+	3	1	4	5	7	5	1
S0)S1)P1)+	2	4	1	5	6	1	5
S0)S1)P1)+	3	1	4	5	6	1	5
S0)S0)S1)P0)+	2	4	4	5	7	5	1
S0)S0)S1)P1)+	2	4	4	5	6	1	5
S0)S1)S1)P0)+	3	4	4	5	7	5	1
S0)S1)S1)P1)+	3	4	4	5	6	1	5
P0)P1)-	6	1	1	2	4	5	1
P0)P1)-	6	1	1	3	4	5	1
P0)P1)-	7	1	1	2	4	1	5
P0)P1)-	7	1	1	3	4	1	5
S0)P0)P1)-	6	1	4	2	4	5	1
S0)P0)P1)-	7	1	4	2	4	1	5
S1)P0)P1)-	6	4	1	3	4	5	1
S1)P0)P1)-	7	4	1	3	4	1	5
P0)P0)P1)-	7	1	1	2	4	5	5
P0)P0)P1)-	7	1	1	3	4	5	5
P0)P1)P1)-	6	1	1	2	4	5	5
P0)P1)P1)-	6	1	1	3	4	5	5
S0)P0)P0)P1)-	7	1	4	2	4	5	5
S0)P0)P1)P1)-	6	1	4	2	4	5	5
S1)P0)P0)P1)-	7	4	1	3	4	5	5
S1)P0)P1)P1)-	6	4	1	3	4	5	5

Denominator terms

b_0							
+	0	1	1	2	4	1	1
+	0	1	1	2	6	1	1
+	0	1	1	2	7	1	1
+	0	1	1	3	4	1	1
+	0	1	1	3	6	1	1
+	0	1	1	3	7	1	1
+	0	1	1	5	6	1	1
+	0	1	1	5	7	1	1
b_1							
S0)+	0	1	4	2	6	1	1
S0)+	0	1	4	2	7	1	1
S0)+	0	1	4	5	6	1	1
S0)+	0	1	4	5	7	1	1
S0)+	2	0	1	2	4	1	1
S0)+	2	0	1	2	6	1	1
S0)+	2	0	1	2	7	1	1
S0)+	2	0	1	3	4	1	1
S0)+	2	0	1	3	6	1	1
S0)+	2	0	1	3	7	1	1
S0)+	2	0	1	5	6	1	1
S0)+	2	0	1	5	7	1	1
b_2							
S1)+	0	4	1	3	4	1	1
S1)+	0	4	1	3	6	1	1
S1)+	0	4	1	3	7	1	1
S1)+	0	4	1	5	6	1	1
S1)+	0	4	1	5	7	1	1
S1)+	3	1	0	2	4	1	1
S1)+	3	1	0	2	6	1	1
S1)+	3	1	0	2	7	1	1
S1)+	3	1	0	3	4	1	1
S1)+	3	1	0	3	6	1	1
S1)+	3	1	0	3	7	1	1
S1)+	3	1	0	5	6	1	1
S1)+	3	1	0	5	7	1	1
b_3							
P0)+	0	1	1	2	4	5	1
P0)+	0	1	1	2	7	5	1
P0)+	0	1	1	3	4	5	1
P0)+	0	1	1	3	7	5	1
P0)+	0	1	1	5	7	5	1

TABLE 3.7 (continued)

Denominator terms

P0)+	7	1	1	2	4	1	0
P0)+	7	1	1	2	6	1	0
P0)+	7	1	1	2	7	1	0
P0)+	7	1	1	3	4	1	0
P0)+	7	1	1	3	6	1	0
P0)+	7	1	1	3	7	1	0
P0)+	7	1	1	5	6	1	0
P0)+	7	1	1	5	7	1	0

b_4

P1)+	0	1	1	2	4	1	5
P1)+	0	1	1	2	6	1	5
P1)+	0	1	1	3	4	1	5
P1)+	0	1	1	3	6	1	5
P1)+	0	1	1	5	6	1	5
P1)+	6	1	1	2	4	0	1
P1)+	6	1	1	2	6	0	1
P1)+	6	1	1	2	7	0	1
P1)+	6	1	1	3	4	0	1
P1)+	6	1	1	3	6	0	1
P1)+	6	1	1	3	7	0	1
P1)+	6	1	1	5	6	0	1
P1)+	6	1	1	5	7	0	1

b_5

S0)S0)+	2	0	4	2	4	1	1
S0)S0)+	2	0	4	2	6	1	1
S0)S0)+	2	0	4	2	7	1	1
S0)S0)+	2	0	4	5	6	1	1
S0)S0)+	2	0	4	5	7	1	1

b_6

S0)S1)+	0	4	4	5	6	1	1
S0)S1)+	0	4	4	5	7	1	1
S0)S1)+	3	0	4	2	4	1	1
S0)S1)+	3	0	4	2	7	1	1
S0)S1)+	2	4	0	3	4	1	1
S0)S1)+	2	4	0	3	6	1	1
S0)S1)+	2	4	0	3	7	1	1
S0)S1)+	2	4	1	0	4	1	1
S0)S1)+	2	4	1	0	6	1	1
S0)S1)+	2	4	1	0	7	1	1
S0)S1)+	3	1	4	0	4	1	1
S0)S1)+	3	1	4	0	6	1	1
S0)S1)+	3	1	4	0	7	1	1
S0)S1)+	2	4	1	5	0	1	1
S0)S1)+	3	1	4	5	0	1	1
S0)S1)+	2	4	1	5	6	0	1
S0)S1)+	3	1	4	5	6	0	1
S0)S1)+	2	4	1	5	7	1	0
S0)S1)+	3	1	4	5	7	1	0

b_7

S0)P0)+	0	1	4	2	4	5	1
S0)P0)+	0	1	4	2	7	5	1
S0)P0)+	0	1	4	5	7	5	1
S0)P0)+	2	0	1	2	4	5	1
S0)P0)+	2	0	1	2	7	5	1
S0)P0)+	2	0	1	3	4	5	1
S0)P0)+	2	0	1	3	7	5	1
S0)P0)+	2	0	1	5	7	5	1
S0)P0)+	7	1	4	2	4	1	0
S0)P0)+	7	1	4	2	6	1	0
S0)P0)+	7	1	4	2	7	1	0
S0)P0)+	7	1	4	5	6	1	0
S0)P0)+	7	1	4	5	7	1	0

b_8

S0)P1)+	0	1	4	2	4	1	5
S0)P1)+	0	1	4	2	6	1	5
S0)P1)+	0	1	4	5	6	1	5
S0)P1)+	2	0	1	2	4	1	5
S0)P1)+	2	0	1	2	6	1	5
S0)P1)+	2	0	1	3	4	1	5
S0)P1)+	2	0	1	3	6	1	5
S0)P1)+	2	0	1	5	6	1	5
S0)P1)+	6	1	4	2	4	0	1
S0)P1)+	6	1	4	2	6	0	1
S0)P1)+	6	1	4	2	7	0	1
S0)P1)+	6	1	4	5	6	0	1
S0)P1)+	6	1	4	5	7	0	1

b_9

S1)S1)+	3	4	0	3	4	1	1
S1)S1)+	3	4	0	3	6	1	1
S1)S1)+	3	4	0	3	7	1	1
S1)S1)+	3	4	0	5	6	1	1
S1)S1)+	3	4	0	5	7	1	1

b_{10}

S1)P0)+	0	4	1	3	4	5	1
S1)P0)+	0	4	1	3	7	5	1
S1)P0)+	0	4	1	5	7	5	1
S1)P0)+	3	1	0	2	4	5	1
S1)P0)+	3	1	0	2	7	5	1

TABLE 3.7 (continued)

Denominator terms

S1)P0)+	3	1	0	3	4	5	1
S1)P0)+	3	1	0	3	7	5	1
S1)P0)+	3	1	0	5	7	5	1
S1)P0)+	7	4	1	3	4	1	0
S1)P0)+	7	4	1	3	6	1	0
S1)P0)+	7	4	1	3	7	1	0
S1)P0)+	7	4	1	5	6	1	0
S1)P0)+	7	4	1	5	7	1	0

b_{11}

S1)P1)+	0	4	1	3	4	1	5
S1)P1)+	0	4	1	3	6	1	5
S1)P1)+	0	4	1	5	6	1	5
S1)P1)+	3	1	0	2	4	1	5
S1)P1)+	3	1	0	2	6	1	5
S1)P1)+	3	1	0	3	4	1	5
S1)P1)+	3	1	0	3	6	1	5
S1)P1)+	3	1	0	5	6	1	5
S1)P1)+	6	4	1	3	4	0	1
S1)P1)+	6	4	1	3	6	0	1
S1)P1)+	6	4	1	3	7	0	1
S1)P1)+	6	4	1	5	6	0	1
S1)P1)+	6	4	1	5	7	0	1

b_{12}

P0)P0)+	7	1	1	2	4	5	0
P0)P0)+	7	1	1	2	7	5	0
P0)P0)+	7	1	1	3	4	5	0
P0)P0)+	7	1	1	3	7	5	0
P0)P0)+	7	1	1	5	7	5	0

b_{13}

P0)P1)+	0	1	1	2	4	5	5
P0)P1)+	0	1	1	3	4	5	5
P0)P1)+	6	0	1	2	4	5	1
P0)P1)+	7	0	1	2	4	1	5
P0)P1)+	6	1	0	3	4	5	1
P0)P1)+	7	1	0	3	4	1	5
P0)P1)+	6	1	1	0	4	5	1
P0)P1)+	7	1	1	0	4	1	5
P0)P1)+	6	1	1	2	0	5	1
P0)P1)+	6	1	1	3	0	5	1
P0)P1)+	6	1	1	5	0	5	1
P0)P1)+	7	1	1	2	0	1	5
P0)P1)+	7	1	1	3	0	1	5
P0)P1)+	7	1	1	5	0	1	5
P0)P1)+	7	1	1	2	6	0	5
P0)P1)+	7	1	1	3	6	0	5
P0)P1)+	7	1	1	5	6	0	5
P0)P1)+	6	1	1	2	7	5	0
P0)P1)+	6	1	1	3	7	5	0
P0)P1)+	6	1	1	5	7	5	0

b_{14}

P1)P1)+	6	1	1	2	4	0	5
P1)P1)+	6	1	1	2	6	0	5
P1)P1)+	6	1	1	3	4	0	5
P1)P1)+	6	1	1	3	6	0	5
P1)P1)+	6	1	1	5	6	0	5

b_{15}

S0)S0)S1)+	2	4	4	0	4	1	1
S0)S0)S1)+	2	4	4	0	6	1	1
S0)S0)S1)+	2	4	4	0	7	1	1
S0)S0)S1)+	2	4	4	5	0	1	1
S0)S0)S1)+	2	4	4	5	6	0	1
S0)S0)S1)+	2	4	4	5	7	1	0

b_{16}

S0)S0)P0)+	2	0	4	2	4	5	1
S0)S0)P0)+	2	0	4	2	7	5	1
S0)S0)P0)+	2	0	4	5	7	5	1

b_{17}

S0)S0)P1)+	2	0	4	2	4	1	5
S0)S0)P1)+	2	0	4	2	6	1	5
S0)S0)P1)+	2	0	4	5	6	1	5

b_{18}

S0)S1)S1)+	3	4	4	0	4	1	1
S0)S1)S1)+	3	4	4	0	6	1	1
S0)S1)S1)+	3	4	4	0	7	1	1
S0)S1)S1)+	3	4	4	5	0	1	1
S0)S1)S1)+	3	4	4	5	6	0	1
S0)S1)S1)+	3	4	4	5	7	1	0

b_{19}

S0)S1)P0)+	0	4	4	5	7	5	1
S0)S1)P0)+	3	0	4	2	4	5	1
S0)S1)P0)+	3	0	4	2	7	5	1
S0)S1)P0)+	2	4	0	3	4	5	1
S0)S1)P0)+	2	4	0	3	7	5	1
S0)S1)P0)+	2	4	1	0	4	5	1

TABLE 3.7 (continued)

Denominator terms

S0)S1)P0)+	2	4	1	0	7	5	1
S0)S1)P0)+	3	1	4	0	4	5	1
S0)S1)P0)+	3	1	4	0	7	5	1
S0)S1)P0)+	2	4	1	5	0	5	1
S0)S1)P0)+	3	1	4	5	0	5	1
S0)S1)P0)+	2	4	1	5	7	5	0
S0)S1)P0)+	3	1	4	5	7	5	0
S0)S1)P0)+	7	4	4	5	6	1	0
S0)S1)P0)+	7	4	4	5	7	1	0

b_{20}

S0)S1)P1)+	0	4	4	5	6	1	5
S0)S1)P1)+	3	0	4	2	4	1	5
S0)S1)P1)+	3	0	4	2	6	1	5
S0)S1)P1)+	2	4	0	3	4	1	5
S0)S1)P1)+	2	4	0	3	6	1	5
S0)S1)P1)+	2	4	1	0	4	1	5
S0)S1)P1)+	2	4	1	0	6	1	5
S0)S1)P1)+	3	1	4	0	4	1	5
S0)S1)P1)+	3	1	4	0	6	1	5
S0)S1)P1)+	2	4	1	5	0	1	5
S0)S1)P1)+	3	1	4	5	0	1	5
S0)S1)P1)+	2	4	1	5	6	0	5
S0)S1)P1)+	3	1	4	5	6	0	5
S0)S1)P1)+	6	4	4	5	6	0	1
S0)S1)P1)+	6	4	4	5	7	0	1

b_{21}

S0)P0)P0)+	7	1	4	2	4	5	0
S0)P0)P0)+	7	1	4	2	7	5	0
S0)P0)P0)+	7	1	4	5	7	5	0

b_{22}

S0)P0)P1)	0	1	4	2	4	5	5
S0)P0)P1)+	2	0	1	2	4	5	5
S0)P0)P1)+	2	0	1	3	4	5	5
S0)P0)P1)+	6	0	4	2	4	5	1
S0)P0)P1)+	7	0	4	2	4	1	5
S0)P0)P1)+	6	1	4	0	4	5	1
S0)P0)P1)+	7	1	4	0	4	1	5
S0)P0)P1)+	6	1	4	2	0	5	1
S0)P0)P1)+	6	1	4	5	0	5	1
S0)P0)P1)+	7	1	4	2	0	1	5
S0)P0)P1)+	7	1	4	5	0	1	5
S0)P0)P1)+	7	1	4	2	6	0	5
S0)P0)P1)+	7	1	4	5	6	0	5
S0)P0)P1)+	6	1	4	2	7	5	0
S0)P0)P1)+	6	1	4	5	7	5	0

b_{23}

S0)P1)P1)+	6	1	4	2	4	0	5
S0)P1)P1)+	6	1	4	2	6	0	5
S0)P1)P1)+	6	1	4	5	6	0	5

b_{24}

S1)S1)P0)+	3	4	0	3	4	5	1
S1)S1)P0)+	3	4	0	3	7	5	1
S1)S1)P0)+	3	4	0	5	7	5	1

b_{25}

S1)S1)P1)+	3	4	0	3	4	1	5
S1)S1)P1)+	3	4	0	3	6	1	5
S1)S1)P1)+	3	4	0	5	4	1	5

b_{26}

S1)P0)P0)+	7	4	1	3	4	5	0
S1)P0)P0)+	7	4	1	3	7	5	0
S1)P0)P0)+	7	4	1	5	7	5	0

b_{27}

S1)P0)P1)+	0	4	1	3	4	5	5
S1)P0)P1)+	3	1	0	2	4	5	5
S1)P0)P1)+	3	1	0	3	4	5	5
S1)P0)P1)+	6	4	0	3	4	5	1
S1)P0)P1)+	7	4	0	3	4	1	5
S1)P0)P1)+	6	4	1	0	4	5	1
S1)P0)P1)+	7	4	1	0	4	1	5
S1)P0)P1)+	6	4	1	3	0	5	1
S1)P0)P1)+	6	4	1	5	0	5	1
S1)P0)P1)+	7	4	1	3	0	1	5
S1)P0)P1)+	7	4	1	5	0	1	5
S1)P0)P1)+	7	4	1	3	6	0	5
S1)P0)P1)+	7	4	1	5	6	0	5
S1)P0)P1)+	6	4	1	3	7	5	0
S1)P0)P1)+	6	4	1	5	7	5	0

b_{28}

S1)P1)P1)+	6	4	1	3	4	0	5
S1)P1)P1)+	6	4	1	3	6	0	5
S1)P1)P1)+	6	4	1	5	6	0	5

TABLE 3.7 (continued)

Denominator terms							
	b_{29}						
P0)P0)P1)+	7	0	1	2	4	5	5
P0)P0)P1)+	7	1	0	3	4	5	5
P0)P0)P1)+	7	1	1	0	4	5	5
P0)P0)P1)+	7	1	1	2	0	5	5
P0)P0)P1)+	7	1	1	3	0	5	5
P0)P0)P1)+	7	1	1	5	0	5	5
	b_{30}						
P0)P1)P1)+	6	0	1	2	4	5	5
P0)P1)P1)+	6	1	0	3	4	5	5
P0)P1)P1)+	6	1	1	0	4	5	5
P0)P1)P1)+	6	1	1	2	0	5	5
P0)P1)P1)+	6	1	1	3	0	5	5
P0)P1)P1)+	6	1	1	5	0	5	5
	b_{31}						
S0)S0)S1)P0)+	2	4	4	0	4	5	1
S0)S0)S1)P0)+	2	4	4	0	7	5	1
S0)S0)S1)P0)+	2	4	4	5	0	5	1
S0)S0)S1)P0)+	2	4	4	5	7	5	0
	b_{32}						
S0)S0)S1)P1)+	2	4	4	0	4	1	5
S0)S0)S1)P1)+	2	4	4	0	6	1	5
S0)S0)S1)P1)+	2	4	4	5	0	1	5
S0)S0)S1)P1)+	2	4	4	5	6	0	5
	b_{33}						
S0)S0)P0)P1)+	2	0	4	2	4	5	5
	b_{34}						
S0)S1)S1)P0)+	3	4	4	0	4	5	1
S0)S1)S1)P0)+	3	4	4	0	7	5	1
S0)S1)S1)P0)+	3	4	4	5	0	5	1
S0)S1)S1)P0)+	3	4	4	5	7	5	0
	b_{35}						
S0)S1)S1)P1)+	3	4	4	0	4	1	5
S0)S1)S1)P1)+	3	4	4	0	6	1	5
S0)S1)S1)P1)+	3	4	4	5	0	1	5
S0)S1)S1)P1)+	3	4	4	5	6	0	5
	b_{36}						
S0)S1)P0)P0)+	7	4	4	5	7	5	0
	b_{37}						
S0)S1)P0)P1)+	3	0	4	2	4	5	5
S0)S1)P0)P1)+	2	4	0	3	4	5	5
S0)S1)P0)P1)+	2	4	1	0	4	5	5
S0)S1)P0)P1)+	3	1	4	0	4	5	5
S0)S1)P0)P1)+	6	4	4	0	4	5	1
S0)S1)P0)P1)+	7	4	4	0	4	1	5
S0)S1)P0)P1)+	2	4	1	5	0	5	5
S0)S1)P0)P1)+	3	1	4	5	0	5	5
S0)S1)P0)P1)+	6	4	4	5	0	5	1
S0)S1)P0)P1)+	7	4	4	5	0	1	5
S0)S1)P0)P1)+	7	4	4	5	6	0	5
S0)S1)P0)P1)+	6	4	4	5	7	5	0
	b_{38}						
S0)S1)P1)P1)+	6	4	4	5	6	0	5
	b_{39}						
S0)P0)P0)P1)+	7	0	4	2	4	5	5
S0)P0)P0)P1)+	7	1	4	0	4	5	5
S0)P0)P0)P1)+	7	1	4	2	0	5	5
S0)P0)P0)P1)+	7	1	4	5	0	5	5
	b_{40}						
S0)P0)P1)P1)+	6	0	4	2	4	5	5
S0)P0)P1)P1)+	6	1	4	0	4	5	5
S0)P0)P1)P1)+	6	1	4	2	0	5	5
S0)P0)P1)P1)+	6	1	4	5	0	5	5
	b_{41}						
S1)S1)P0)P1)+	3	4	0	3	4	5	5
	b_{42}						
S1)P0)P0)P1)+	7	4	0	3	4	5	5
S1)P0)P0)P1)+	7	4	1	0	4	5	5
S1)P0)P0)P1)+	7	4	1	3	0	5	5
S1)P0)P0)P1)+	7	4	1	5	0	5	5

TABLE 3.7 (continued)

Denominator terms							
b_{43}							
S1)PO)P1)P1)+	6	4	0	3	4	5	5
S1)PO)P1)P1)+	6	4	1	0	4	5	5
S1)PO)P1)P1)+	6	4	1	3	0	5	5
S1)PO)P1)P1)+	6	4	1	5	0	5	5
b_{44}							
SO)SO)S1)PO)P1)+	2	4	4	0	4	5	5
SO)SO)S1)PO)P1)+	2	4	4	5	0	5	5
b_{45}							
SO)S1)S1)PO)P1)+	3	4	4	0	4	5	5
SO)S1)S1)PO)P1)+	3	4	4	5	0	5	5
b_{46}							
SO)S1)PO)PO)P1)+	7	4	4	0	4	5	5
SO)S1)PO)PO)P1)+	7	4	4	5	0	5	5
b_{47}							
SO)S1)PO)P1)P1)+	6	4	4	0	4	5	5
SO)S1)PO)P1)P1)+	6	4	4	5	0	5	5

appear as coefficients for. The a's in Eq. (3.28) are the coefficients of the proper combination of S_0, S_1, P_0, and/or P_1 for the rate equation numerator in Table 3.6, and the b's are the coefficients of the proper combination of these concentrations in Table 3.7.

Equation (3.28) can also be written in the form of Eq. (3.26) for $x = S_0$ or $x = S_1$. When x is S_0 the coefficients in Eq. (3.26) are

$$A_0 = a_{13}P_0P_1 + a_{27}S_1P_0P_1 + a_{29}P_0^2P_1 + a_{30}P_0P_1^2 + a_{42}S_1P_0^2P_1 + a_{43}S_1P_0P_1^2$$

$$A_1 = a_6S_1 + a_{18}S_1^2 + a_{19}S_1P_0 + a_{20}S_1P_1 + a_{22}P_0P_1 + a_{34}S_1^2P_0 + a_{35}S_1^2P_1 + a_{39}P_0^2P_1 + a_{40}P_0P_1^2$$

$$A_2 = a_{15}S_1 + a_{31}S_1P_0 + a_{32}S_1P_1$$

$$\begin{aligned}
B_0 = {} & b_0 + b_2S_1 + b_3P_0 + b_4P_1 + b_9S_1^2 + b_{10}S_1P_0 + b_{11}S_1P_1 \\
& + b_{12}P_0^2 + b_{13}P_0P_1 + b_{14}P_1^2 + b_{24}S_1^2P_0 + b_{25}S_1^2P \\
& + b_{26}S_1P_0^2 + b_{27}S_1P_0P_1 + b_{28}S_1P_1^2 + b_{29}P_0^2P_1 \\
& + b_{30}P_0P_1^2 + b_{41}S_1^2P_0P_1 + b_{42}S_1P_0^2P_1 + b_{43}S_1P_0P_1^2
\end{aligned}$$

$$\begin{aligned}
B_1 = {} & b_1 + b_6S_1 + b_7P_0 + b_8P_1 + b_{18}S_1^2 + b_{19}S_1P_0 \\
& + b_{20}S_1P_1 + b_{21}P_0^2 + b_{22}P_0P_1 + b_{23}P_1^2 + b_{34}S_1^2P_0 \\
& + b_{35}S_1^2P_1 + b_{36}S_1P_0^2 + b_{37}S_1P_0P_1 + b_{38}S_1P_1^2 + b_{39}P_0^2P_1 \\
& + b_{40}P_0P_1^2 + b_{45}S_1^2P_0P_1 + b_{46}S_1P_0^2P_1 + b_{47}S_1P_0P_1^2
\end{aligned}$$

$$\begin{aligned}
B_2 = {} & b_5 + b_{15}S_1 + b_{16}P_0 + b_{17}P_1 \\
& + b_{31}S_1P_0 + b_{32}S_1P_1 + b_{33}P_0P_1 + b_{44}S_1P_0P_1
\end{aligned}$$

When x is S_1 the coefficients are

$$\begin{aligned}
A_0 = {} & a_{13}P_0P_1 + a_{22}S_0P_0P_1 + a_{29}P_0^2P_1 \\
& + a_{30}P_0P_1^2 + a_{39}S_0P_0^2P_1 + a_{40}S_0P_0P_1^2
\end{aligned}$$

$$\begin{aligned}
A_1 = {} & a_6S_0 + a_{15}S_0^2 + a_{19}S_0P_0 + a_{20}S_0P_1 \\
& + a_{27}P_0P_1 + a_{31}S_0^2P_0 + a_{32}S_0^2P_1 + a_{42}P_0^2P_1 + a_{43}P_0P_1
\end{aligned}$$

$$A_2 = a_{18}S_0 + a_{34}S_0P_0 + a_{35}S_0P_1$$

$$\begin{aligned}
B_0 = {} & b_0 + b_1S_0 + b_3P_0 + b_4P_1 + b_5S_0^2 + b_7S_0P_0 + b_8S_0P_1 \\
& + b_{12}P_0^2 + b_{13}P_0P_1 + b_{14}P_1^2 + b_{16}S_0^2P_0 + b_{17}S_0^2P_1 \\
& + b_{21}S_0P_0^2 + b_{22}S_0P_0P_1 + b_{23}S_0P_1^2 + b_{29}P_0^2P_1 \\
& + b_{30}P_0P_1^2 + b_{33}S_0^2P_0P_1 + b_{39}S_0P_0^2P_1 + b_{40}S_0P_0P_1^2
\end{aligned}$$

$$B_1 = b_2 + b_6S_0 + b_{10}P_0 + b_{11}P_1 + b_{15}S_0^2 + b_{19}S_0P_0$$
$$+ b_{20}S_0P_1 + b_{26}P_0^2 + b_{27}P_0P_1 + b_{28}P_1^2 + b_{31}S_0^2P_0$$
$$+ b_{32}S_0^2P_1 + b_{36}S_0P_0^2 + b_{37}S_0P_0P_1 + b_{38}S_0P_1^2 + b_{42}P_0^2P_1$$
$$+ b_{43}P_0P_1^2 + b_{44}S_0^2P_0P_1 + b_{46}S_0P_0^2P_1 + b_{47}S_0P_0P_1^2$$

$$B_2 = b_9 + b_{18}S_0 + b_{24}P_0 + b_{25}P_1$$
$$+ b_{34}S_0P_0 + b_{35}S_0P_1 + b_{41}P_0P_1 + b_{45}S_0P_0P_1$$

3.3 INTEGRATED RATE EQUATIONS FOR QUASI-STEADY-STATE ENZYME MECHANISMS

Quasi-steady-state rate equations like Eqs. (3.17) and (3.26) are *differential* equations that relate $\dot{P}$ to S and P. Strictly speaking there are fixed, quasi-steady-state concentrations of substrate [S(t*)] and product [P(t*)]. If these quantities are substituted into quasi-steady-state rate equations, $\dot{P}(t^*)$ also assumes a fixed, constant value. However, during *short* time intervals, it is often permissible to consider the variations in S(t) and P(t) without discarding the approximation that the concentrations of the enzyme intermediates are quasi-stationary. Furthermore, since it is usually S(t) or P(t) rather than $\dot{P}$ that is the experimentally measured quantity, it is natural to seek the *integrals* of quasi-steady-state rate equations.

In closed systems the total amount of matter added initially is conserved. If the initial enzyme concentration is low compared to the concentration of substrate added initially, then the amount of substrate or product bound to the enzyme must also be low. In general, when the bound fractions of substrate and product can be neglected in overall reactions with stoichiometries of one,

$$S_i(0) = S_i(t) + P(t) \qquad [i = 0, \ldots, n] \qquad (3.29)$$

where

$$P(t) \equiv P_i(t) \qquad [i = 0, \ldots, n] \tag{3.30}$$

If we use $S = S(0) - P$ to eliminate S from Eq. (3.17), we obtain

$$\dot{P} = \frac{\xi_0 + \xi_1 P}{\zeta_0 + \zeta_1 P} \tag{3.31}$$

For the mechanisms described by Eqs. (3.19) or (3.27) we use $S_0 = S_0(0) - P$ and $S_1 = S_1(0) - P$ to eliminate S_0 and S_1 from Eq. (3.26). For the mechanism described by Eqs. (3.19),

$$\dot{P} = \frac{\xi_0 + \xi_1 P + \xi_2 P^2 + \xi_3 P^3}{\zeta_0 + \zeta_1 P + \zeta_2 P^2 + \zeta_3 P^3} \tag{3.32}$$

and for the mechanism described by Eqs. (3.27),

$$\dot{P} = \frac{\xi_0 + \xi_1 P + \xi_2 P^2 + \xi_3 P^3 + \xi_4 P^4 + \xi_5 P^5}{\zeta_0 + \zeta_1 P + \zeta_2 P^2 + \zeta_3 P^3 + \zeta_4 P^4 + \zeta_5 P^5} \tag{3.33}$$

where the ξ's and ζ's are constants that depend on the initial conditions and the rate constants in each mechanism. In general, quasi-steady-state rate equations take the form of a rational function of degree q,

$$\dot{P} = \frac{\xi_0 + \sum_{i=1}^{p} \xi_i P^i}{\zeta_0 + \sum_{i=1}^{q} \zeta_i P^i} \tag{3.34}$$

where P refers to the concentration of any of the various products. Equation (3.34) is a first-order differential equation involving one variable. To obtain the analytic solution for the finite interval $t_1 \simeq 0 \leq t \leq t_2$, we must find the integral of

$$\int_0^{t_2} dt = \int_{P(0)}^{P(t_2)} \frac{\zeta_0 + \sum_{i=1}^{q} \zeta_i P^i}{\xi_0 + \sum_{i=1}^{p} \xi_i P^i} \, dP \equiv \int_{P(0)}^{P(t_2)} F(P) \, dP \tag{3.35}$$

Since F(P) is a rational function of P, the finite interval P(0) to $P(t_2)$ can be chosen such that it can be represented by a polynomial in P:

$$F(P) = \sum_{i=0}^{n} \eta_i P^i \tag{3.36}$$

where

$$\eta_0 = F(0) = \frac{\zeta_0}{\xi_0} \tag{3.37}$$

$$\eta_1 = F'(0) = \frac{\xi_0\zeta_1 - \xi_1\zeta_0}{\xi_0^2} \tag{3.38}$$

and

$$\eta_i = \frac{F^i(0)}{(i - 1)!} \qquad [i = 2, \ldots, n] \tag{3.39}$$

Since F(P) is a polynomial of power n, the integral of F(P) must be a polynomial of power n + 1:

$$\int_{P(0)}^{P(t_2)} F(P)\, dP = C + \sum_{i=1}^{n+1} \frac{\eta_{i-1}}{i} P^i = t \tag{3.40}$$

C in Eq. (3.40) is a constant of integration [for P(0) = 0, C ≃ 0], and n + 1 is the highest power of P necessary for the representation of P(t) in the interval $0 \leq t \leq t_2$. For many enzyme reactions, the quadratic representation

$$C + \frac{\zeta_0}{\xi_0} P + \frac{\xi_0\zeta_1 - \xi_1\zeta_0}{\xi_0^2} P^2 = t \tag{3.41}$$

is adequate to describe the quasi-steady-state time course during most of the far-from-equilibrium [P(t)/S(t) << P(0)/S(0)] stage of the overall reaction.[11]

3.4 HOMEOMORPHISM

When two different models lead to the same mathematical description, the models are said to be *homeomorphic*. Since unconsidered models might be homeomorphic with the currently popular model, one can never logically assert that a particular model is surely correct. Models can be logically rejected because they are inconsistent with observations, but no model can be established as the *only* possible model. Thus, homeomorphism is the logical basis for universal scientific agnosticism.

Complex models described by exact mathematical relationships like Eq. (3.1) often possess subtle features that distinguish them from a large number of other models. But when *approximate* mathematical relationships like inequality (3.2) can be used to describe models, the distinguishing features of the exact relationships may disappear. Thus the use of approximate models like inequality (3.2) can lead to an overlap of the behavior of otherwise different models. The practical result is that models that can be distinguished in principle cannot be distinguished in practice.

When two different models lead to overlaping z in inequality (3.2), the models are said to be *approximately homeomorphic*. For *valid* approximations, approximately homeomorphic mechanisms are as indistinguishable as though they were homeomorphic.

The quasi-steady-state assumption is a good example of an approximation that is often sufficiently valid to cause different enzyme mechanisms to become approximately homeomorphic. For example, the two mechanisms described by Eqs. (3.3) and (3.9) are different, but they become approximately homeomorphic when the quasi-steady-state assumption is valid. Although there are six parameters in Eqs. (3.3) and eight in Eqs. (3.9), a quasi-steady-state description of either mechanism leads to the calculation of only four parameters: V_f, V_r, K_f, and K_r defined after Eqs. (1.18) and (1.19). The rest of the information theoretically available from the kinetic behavior of these two mechanisms, and the ability to distinguish between them is lost in the approximation.

It is also possible for x* in Eq. (3.1) to be in the range of z in inequality (3.2). When the exact solution for one model falls in the range of the approximate solution for another model, the models are said to be *nearly homeomorphic*. Again, for *valid* approximations, nearly homeomorphic mechanisms are as indistinguishable as though they were homeomorphic.

Identical mathematical relationships can arise for non-steady-state models of simple enzyme mechanisms (for example, mechanisms involving one substrate or ordered sequence mechanisms) and quasi-steady-state mechanisms of complex reactions (for example, random-order mechanisms).

3.5 THE ALLOCATION OF ERROR TO THE MODEL AND TO EXPERIMENT

The principal method of deciding whether or not to reject a model is to compare the expected and the experimental behaviors. The test is usually accomplished by taking into account experimental error and estimating how likely the experimental results are *if the model is perfectly valid*. A common measure of the deviation between a perfectly valid model x* and its measurement $\bar{x}$ is

$$D(x^*) = \int [x^*(t) - \bar{x}(t)]^2 \, dt \tag{3.42}$$

for continuous data, or

$$\sum [x^*(t_i) - \bar{x}(t_i)]^2 \tag{3.43}$$

for descrete data sets $[x(t_i), t_i]$. If the model contains parameters, these definitions are modified to make the deviations minimum with respect to the parameters. The resulting D(x*) is the usual least-square deviation, and x* specified by the estimated parameters is called the "best fit." The decision about whether to reject a model that has been obtained in this way is usually made by comparing the estimated probability distribution of an assumed random error in the measurement and the calculated probability that a deviation will exceed $D(x^*)$ when the error is imposed on x^*. If this probability is reasonably large, then the model is retained as a possible model; if the probability is small, the model is said to be unlikely.

If models were always without error, there would be no near homeomorphism and no approximate homeomorphism; there would only be true homeomorphism. Models for which simplifying assumptions have been made obviously contain error and must be described by inequality (3.2) rather than Eq. (3.1). However, even models for which no simplifying assumptions have been made are mere guesses, and they should also be expected to contain error.[11]

A method that provides for the allocation to theory and experiment any discrepancy between them and therefore takes explicit account of error in models had been developed.[12] In the composite deviation,

$$C(z) = w\,\overline{M}(z) + D(z) \tag{3.44}$$

where

$$\overline{M}(z) = \int [M(\dot{z},z,t)]^2\,dt$$

and w is a weighting factor that reflects the relative importance of the deviations due to errors in the model $[M(z)]$ and the measurements $[D(z)]$. Equation (3.44) therefore provides a measure of how well the model is satisfied by the data *and* how well the measurement is approximated.

The minimum composite deviation $C(z^*)$ is obtained by minimizing Eq. (3.44) with respect to the various parameters in the model. The mathematical procedure is a standard problem in the calculus of variations, and it leads to the Euler-Lagrange equation and the supplementary boundary conditions for the minimum function z^*. For the general approximate enzyme model described by Eq. (3.34),

$$M(P,P) = P - \frac{\xi_0 + \sum_{i=1}^{p} \xi_i P^i}{\zeta_0 + \sum_{i=1}^{q} \zeta_i P^i} \equiv \dot{P} - F(P) \tag{3.45}$$

P represents the quasi-steady-state range of values for the product concentration. The function P* yielding the minimum composite deviation is given by the Euler-Lagrange equation when P = P*:

$$w\ddot{P} - w\,F(P)\,\frac{\partial F(P)}{\partial P} - (P - \bar{P}) = 0 \tag{3.46}$$

and the natural boundary conditions

$$\dot{P} - F(P) = 0 \tag{3.47}$$

and

$$P = P(t_1),\ P(t_2) \tag{3.48}$$

$\bar{P}$ in Eq. (3.46) is the experimentally determined concentration, and $P(t_1)$ and $P(t_2)$ in Eq. (3.48) are the initial and final values of P during which the quasi-steady-state approximation is used.

As w in Eq. (3.44) becomes small, the best-fitting solution z* tends to the measurement $\bar{x}$, but as w becomes large, z* tends toward x*. We expect therefore that a maximum allowable value for D(z*) will define an upper bound for w, and a maximum allowable value for ε will define a lower bound for w. If the upper bound exceeds the lower bound, there is a z* known in advance that satisfies the model *and* approximates the measurement. However, if the lower bound for w exceeds the upper bound, there is no function simultaneously satisfying the advance conditions of the model and the measurement errors. In either case each bound separately limits the "validity" of the model relative to the experimental measurements.

Quasi-steady-state enzyme kinetics was developed for the practical purpose of studying chemical models of enzyme action. It turns out, however, that most enzyme mechanisms cannot be distinguished from a large group of other mechanisms unless one knows in advance the extent to which the quasi-steady-state approximation is valid and the extent to which the mechanisms are nearly homeomorphic. Furthermore, even if one knows in advance that the quasi-steady-state assumption is a good approximation for a particular enzyme, a large number of such mechanisms are approximately homeomorphic. One of the unfortunate aspects of the development of "theoretical" quasi-steady-state enzyme kinetics is that the "theoreticians" have not been sufficiently concerned with near homeomorphism and approximate homeomorphism and other aspects of model error.[2]

CHAPTER 4
SIMULATIONS OF ENZYME MODELS

4.1 SIMULATIONS ON ELECTRONIC ANALOG COMPUTERS

The most straightforward way to ascertain the mathematical relationships expected for a model is to assume certain physical laws govern the components in the model, write down the associated mathematical description for each component, and solve the system of equations. Usually, however, one finds that the system of equations cannot be solved analytically. In such cases it is common practice to assume that specific approximate relationships describe the behavior of the components. Eventually the equations become sufficiently simplified that they can be solved.

An alternative to making mathematical approximations about the relationships that describe the components in a model is to construct an analog of the model. The analog could take any physical form, but the most convenient and accurate analog machine commercially available is the electronic analog computer. Electronic analog computers possess devices that simulate addition, subtraction, multiplication, division, integration, and other mathematical operations.

Even very accurate electronic analog computers do not simulate models *exactly*. As a consequence, simulations made in the neighborhood of instabilities can contain extremely large errors due to the originally small fluxuations near the instability. But most accurate electronic analog computer simulations of enzyme mechanisms with no instabilities should be at least as accurate as the approximate mathematical models derived on the basis of the quasi-steady-state assumption.

In order to use electronic analog computers, one must know how to program them. The conventions for the generation of a constant, multiplication or division by a constant, inversion of sign, addition or substraction, integration and initial conditions, and multiplication or division of two variables used in electronic-analog-computer program diagrams appear in Table 4.1.

The enzyme mechanism in Eqs. (3.3) is described by five differential equations,

$$\dot{S} = k_{-1}X_1 - k_1X_0S \tag{4.1}$$

$$\dot{X}_0 = k_{-1}X_1 + k_3X_2 - k_1X_0S - k_{-3}X_0P \tag{4.2}$$

$$\dot{X}_1 = k_1X_0S - k_{-2}X_2 - k_{-1}X_1 - k_2X_1 \tag{4.3}$$

$$\dot{X}_2 = k_2X_1 + k_{-3}X_0P - k_{-2}X_2 - k_3X_2 \tag{4.4}$$

$$\dot{P} = k_3X_2 - k_{-3}X_0P \tag{4.5}$$

and two algebraic equations,

$$E(0) = X_0 + X_1 + X_2 \tag{4.6}$$

$$S(0) = S + P + X_1 + X_2 \tag{4.7}$$

Since there are only five independent variables, we need only five of these seven equations. Equation (4.7) is not a good one to use in analog simulations because S and P are often considerably larger than X_1 and X_2, and addition would tend to simulate $S(0) = S + P$ rather than Eq. (4.7). Equation (4.6) does not possess this disadvantage so we select it, and for the other four equations, we choose Eqs. (4.1), (4.3), (4.4), and (4.5).

The next step is to select the proper scale factor for the

TABLE 4.1. Some conventions used in electronic analog computer programs in this chapter.

Symbol[a]	Operation
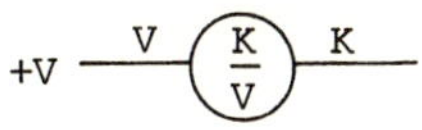	Generation of a constant
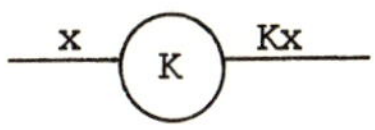	Multiplication by a constant ≤ 1
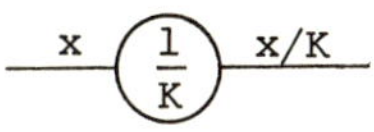	Division by a constant ≥ 1
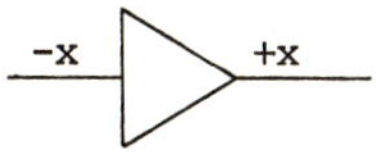	Inversion of the sign of a variable
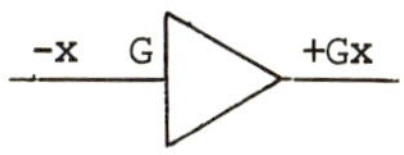	Inversion with gain G
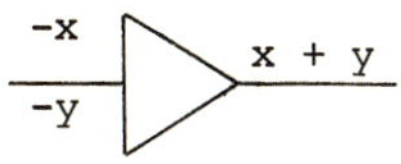	Addition of two variables
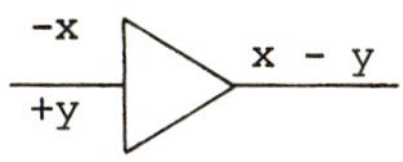	Subtraction of two variables
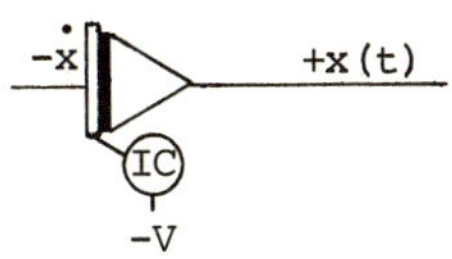	Integration
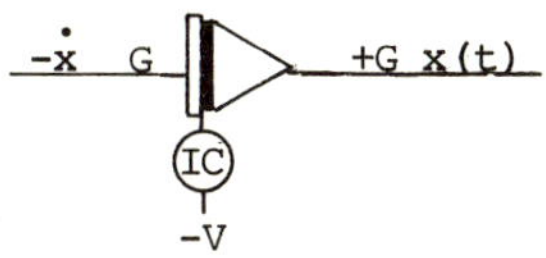	Integration with gain G
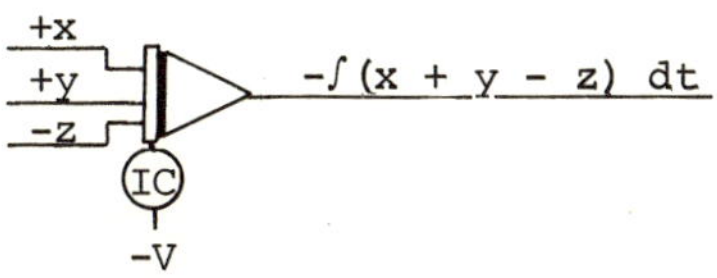	Addition, subtraction, and integration

TABLE 4.1 (continued)

Symbol[a]	Operation
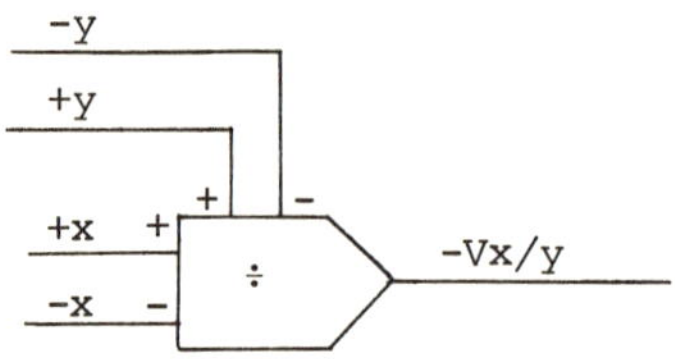	Division of two variables
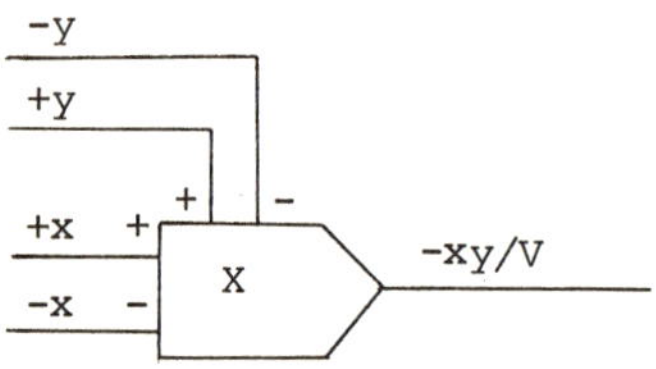	Multiplication of two variables
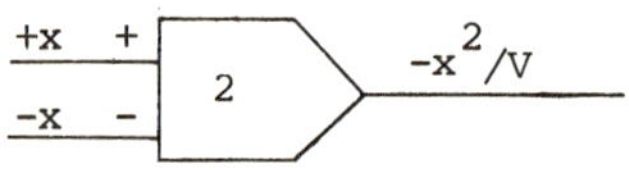	Square

[a]V is the maximum voltage of the analog machine used.

concentrations of the five variables. The best procedure is to identify the maximum concentration that any variable can have, divide this maximum concentration by the maximum voltage permissible on the analog machine, and obtain a scale factor in concentration units per volt. For example, suppose that in Eqs. (4.6) and (4.7) $E(0) = 1\ \mu M$, and $S(0) = 100\ \mu M$. Then none of the concentrations can exceed $100\ \mu M$, and we obtain a scale facter of $10\ \mu M/V$ for a 10-V analog machine and $1\ \mu M/V$ for a 100-V computer. We now obtain the *scaled variables*, $s = 10^5 S$, $p = 10^5 P$, $x_0 = 10^5 X_0$, $x_1 = 10^5 X_1$, and $x_2 = 10^5 X_2$, and the *scaled constants*, $K_1 = 10^{-5} k_1$, $K_{-1} = k_{-1}$, $K_2 = k_2$, $K_{-2} = k_{-2}$, $K_3 = k_3$, and $k_{-3} = 10^{-5} k_{-3}$ for a 10-V machine. The units of s, p, x_0, x_1, and x_2 are volts, the units of K_{-1}, K_2, K_{-2}, and K_3 are reciprocal time (arbitrary units), and the units of K_1 and K_{-3} are reciprocal volt-time. The *scaled equations* are:

$$\dot{s} = K_{-1}x_1 - K_1 x_0 s \tag{4.8}$$

$$\dot{x}_1 = K_1 x_0 s + K_{-2}x_2 - K_{-1}x_1 - K_2 x_1 \tag{4.9}$$

$$\dot{x}_2 = K_2 x_1 + K_{-3} x_0 p - K_{-2} x_2 - K_3 x_2 \tag{4.10}$$

$$\dot{p} = K_3 x_2 - K_{-3} x_0 p \tag{4.11}$$

$$e(0) = x_0 + x_1 + x_2 \tag{4.12}$$

The electronic-analog-computer diagram for these equations and a 10-V machine is illustrated in Fig. 4.1.

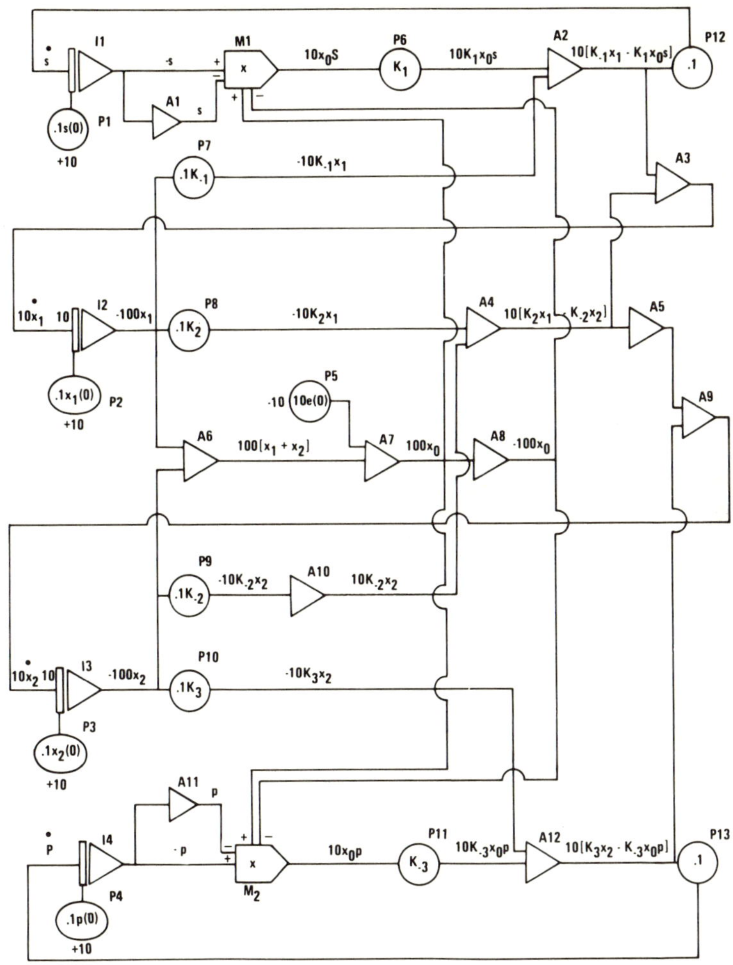

FIG. 4.1. Analog-computer diagram for the enzyme mechanism described by Eqs. (4.8)-(4.12).

The analog components used to simulate multiplication by a constant and the initial conditions in Fig. 4.1 are variable potentiometers, and the components used to achieve inversion of sign, summation or subtraction, and integration are amplifiers. In this figure integrator I1 integrates Eq. (4.8), I2 integrates Eq. (4.9) with a gain of 10, I3 integrates Eq. (4.10) with a gain of 10, and I4 integrates Eq. (4.11). The potentiometer setting for P1 is 0.1 s(0), P2 is 0.1 $x_1(0)$, P3 is 0.1 $x_2(0)$, P4 is 0.1 p(0) and P5 is 10 e(0). P6 is K_1, P7 is $0.1K_{-1}$, P8 is $0.1K_2$, P9 is $0.1K_{-2}$, P10 is $0.1K_3$, and P11 is K_{-3}. P12 and P13 are scaling potentiometers permanently set to 0.1. Multiplier M1 multiplies s times $100x_0$, and multiplier M2 multiplies p times $100x_0$; these products are negated and divided by 10. Amplifier A1 inverts the sign of -s, A2 sums $10K_1x_0s$ and $-10K_{-1}x_1$ and inverts the sign of the difference, A3 sums and inverts $10(K_{-1}x_1 - K_1x_0s)$ plus $10(K_2x_1 - K_{-2}x_2)$, A4 sums and inverts $10K_{-2}x_2$ and $-10K_2x_1$, A5 inverts $10(K_2x_1 - K_{-2}x_2)$, A6 sums and inverts $-100x_1$ plus $-100x_2$, A7 sums and inverts $100(x_1 + x_2)$ plus -100 e(0), A8 inverts $100x_0$, A9 sums and inverts $10(K_{-2}x_2 - K_2x_1)$, A10 inverts $-10K_{-2}x_2$, A11 inverts -P, and A12 sums and inverts $10K_{-3}x_0p$ plus $-10K_3x_2$.

One of the disadvantages of simulation studies is that specific values for the initial conditions and the rate constants must be chosen for each simulation, and one obtains a solution for only one set of parameter values per simulation. In principle, generality can be approached if one obtains simulations for all possible combinations of all possible values of the parameters. What one does in practice, however, is to set the values of all the parameters but one, and then obtain solutions at very high frequency while turning the potentiometer dial for the varying parameter. If this procedure is repeated for a large number of combinations of the unvarying parameters, results that approach generality in the parameter range used may be obtained. This procedure requires an analog machine capable of repetitive, high-speed operation so that slow problems can be scaled to run in a few milliseconds.

It is relatively easy to write analog computer programs for complex enzyme models. For example, the mechanism in Eqs. (3.27) is described by ten differential equations and one algebraic equation:

$$\dot{S}_0 = k_{-1}X_1 + k_3X_2 - k_1X_0S_0 - k_{-3}X_3S_0 \tag{4.13}$$

$$\dot{S}_1 = k_{-2}X_2 + k_4X_3 - k_2X_1S_1 - k_{-4}S_0S_1 \tag{4.14}$$

$$\dot{X}_1 = k_1X_0S_0 + k_{-2}X_2 - k_{-1}X_1 - k_2X_1S_1 \tag{4.15}$$

$$\dot{X}_2 = k_2X_1S_1 + k_{-3}X_3S_0 + k_{-5}X_4 - k_{-2}X_2 - k_3X_2 - k_5X_2 \tag{4.16}$$

$$\dot{X}_3 = k_3X_2 + k_{-4}X_0S_1 - k_{-3}X_3S_0 - k_4X_3 \tag{4.17}$$

$$\dot{X}_4 = k_5X_2 + k_{-6}X_5P_0 + k_9X_6P_1 - k_{-5}X_4 - k_6X_4 - k_{-9}X_4 \tag{4.18}$$

$$\dot{X}_5 = k_6X_4 + k_{-7}X_0P_1 - k_{-6}X_5P_0 - k_7X_5 \tag{4.19}$$

$$\dot{X}_6 = k_8k_0P_0 + k_{-9}X_4 - k_{-8}X_6 - k_9X_6P_1 \tag{4.20}$$

$$\dot{P}_0 = k_6X_4 + k_{-8}X_6 - k_{-6}X_5P_0 - k_8X_0P_0 \tag{4.21}$$

$$\dot{P}_1 = k_7X_5 + k_{-9}X_4 - k_{-7}X_0P_1 - k_9X_6P_1 \tag{4.22}$$

$$E(0) = X_0 + X_1 + X_2 + X_3 + X_4 + X_5 + X_6 \tag{4.23}$$

Using the scale factor 10 μ*M*/V, we obtain the scaled variables, $s_0 = 10^5S_0$, $s_1 = 10^5S_1$, $x_0 = 10^5X_0$, $x_1 = 10^5X_1$, $x_2 = 10^5X_2$, $x_3 = 10^5X_3$, $x_4 = 10^5X_4$, $x_5 = 10^5X_5$, $x_6 = 10^5X_6$, $p_0 = 10^5P_0$, and $p_1 = 10^5P_1$, and the scaled constants, $K_1 = 10^{-5}k_1$, $K_{-1} = k_{-1}$, $K_2 = 10^{-5}k_2$, $K_{-2} = k_{-2}$, $K_3 = k_3$, $K_{-3} = 10^{-5}k_{-3}$, $K_4 = k_4$, $K_{-4} = 10^{-5}k_{-4}$, $K_5 = k_5$, $K_{-5} = k_{-5}$, $K_6 = k_6$, $K_{-6} = 10^{-5}k_{-6}$, $K_7 = k_7$, $K_{-7} = 10^{-5}k_{-7}$, $K_8 = 10^{-5}k_8$, $K_{-8} = k_{-8}$, $K_9 = 10^{-5}k_9$, and $K_{-9} = k_{-9}$ for a 10-V machine. The scaled equations are

$$\dot{s}_0 = K_{-1}x_1 + K_3x_2 - K_1x_0s_0 - K_{-3}x_3s_0 \tag{4.24}$$

$$\dot{s}_1 = K_{-2}x_2 + K_4x_3 - K_2x_1s_1 - K_{-4}x_0s_1 \tag{4.25}$$

$$\dot{x}_1 = K_1 x_0 s_0 + K_{-2} x_2 - K_{-1} x_1 - K_2 x_1 s_1 \tag{4.26}$$

$$\dot{x}_2 = K_2 x_1 s_1 + K_{-3} x_3 s_0 + K_{-5} x_4 - K_{-2} x_2 - K_3 x_2 - K_5 x_2 \tag{4.27}$$

$$\dot{x}_3 = K_3 x_2 + K_{-4} x_0 s_1 - K_{-3} x_3 s_0 - K_4 x_3 \tag{4.28}$$

$$\dot{x}_4 = K_5 x_2 + K_{-6} x_5 p_0 + K_9 x_6 p_1 - K_{-5} x_4 - K_6 x_4 - K_{-9} x_4 \tag{4.29}$$

$$\dot{x}_5 = K_6 x_4 + K_{-7} x_0 p_1 - K_{-6} x_5 p_0 - K_7 x_5 \tag{4.30}$$

$$\dot{x}_6 = K_8 x_0 p_0 + K_{-9} x_4 - K_{-8} x_6 - K_9 x_6 p_1 \tag{4.31}$$

$$e(0) = x_0 + x_1 + x_2 + x_3 + x_4 + x_5 + x_6 \tag{4.32}$$

$$\dot{p}_0 = K_6 x_4 + K_{-8} x_6 - K_{-6} x_5 p_0 - K_8 x_0 p_0 \tag{4.33}$$

$$\dot{p}_1 = K_7 x_5 + K_{-9} x_4 - K_{-7} x_0 p_1 - K_9 x_6 p_1 \tag{4.34}$$

The analog-computer diagram for these equations in illustrated in Figure 4.2.

The program in Figure 4.2 can be used to represent a variety of special mechanisms involving two substrates and two products. For example, an ordered sequence of S_0 addition followed by S_1 addition results if the potentiometers representing K_3, K_{-3}, K_4, and K_{-4} are set to zero. An ordered sequence of P_0 desorption followed by P_1 desorption is obtained if the potentiometers representing K_8, K_{-8}, K_9, and K_{-9} are zero. If the potentiometers representing K_3 and K_{-3} or K_4 and K_{-4} (but not both) are set to zero, the substrate addition is ordered, but inactive enzyme-S_1 or enzyme-S_0 compounds are present. If the potentiometers representing K_8 and K_{-8} or K_9 and K_{-9} are set to zero, the product desorption is ordered, but inactive enzyme-P_1 or enzyme-P_2 compounds are present. Abortive enzyme compounds of this type have been reported for a number of 2S, 2P-ordered-(SSPP) mechanisms.

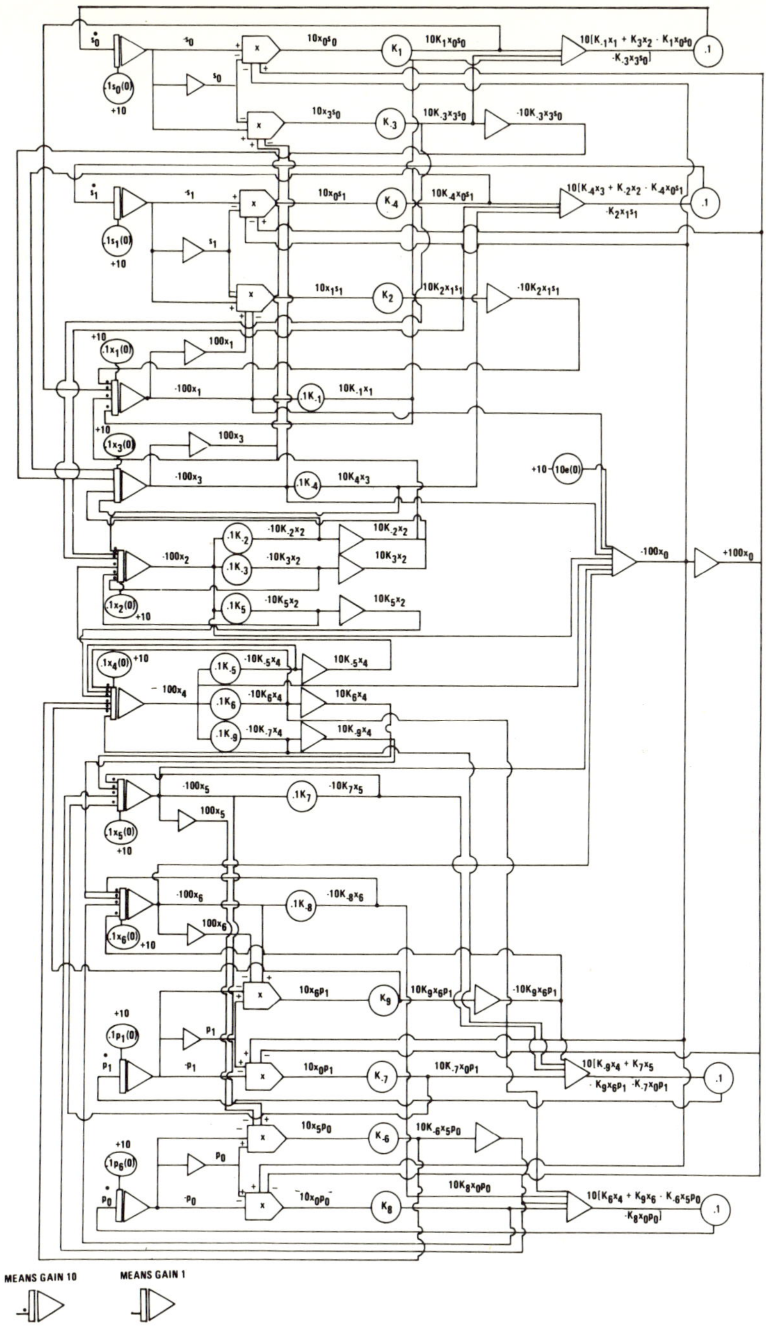

FIG. 4.2. Analog computer diagram for the enzyme mechanism described by Eqs. (4.24)-(4.34).

4.2 NUMERICAL SIMULATIONS ON DIGITAL COMPUTERS

Another alternative to making assumptions about the relationships that govern the components in a model is to obtain numerical solutions on a high-speed digital computer. Numerical solutions are similar to analog simulations in the sense that one obtains only a unique solution for the particular set of constants and initial conditions used. However, a distinct advantage of numerical solutions is that one can be obtained as accurately as desired (to within any stipulated small error) without changing hardware components in the computer. A disadvantage of digital solutions is that complex problems (for example, problems requiring the solution of certain nonlinear partial differential equations) may require considerably more time on a digital computer. It seems likely that future simulation techniques for complex biological models will necessarily combine the advantages of analog computers (speed of integration) and digital computers (accuracy) and use hybrid analog/digital systems.

Fortunately for our purposes, the mathematical relationships describing the enzyme models we have been discussing are a set of *ordinary* differential equations. Notwithstanding the "stiffness" one encounters with these equations, high-speed digital computation with current-generation computers is feasible even for models involving 100 or 200 chemical steps. For example, the "University College London Enzyme Simulator" is a digital-computer program that uses the Gear-Curtis technique for integration of "stiff" ordinary differential equations describing enzyme mechanisms involving up to 200 components.[13] This program is especially easy to use because it translates the individual chemical steps into mathematical relationships via an "Equation Translator Routine." The input is thus a set of elementary chemical reactions together with the values for the various rate constants and initial conditions; the output is a display of the time course of the chemical components.

The Kinetic Translation Language (KTL) of the University College London Enzyme Simulator has three types of input records:

Comments are designated by "*" as the first nonblank character, *chemical reactions* are elementary chemical steps with related initial conditions and one or two rate constants, and *end of KTL list* markers are designated by "**b", (where "b" represents a blank). The comment and chemical-reaction records can be up to 240 characters long, and they are always terminated by "b,b". Reactants in reaction statements are separated by a "+" or "=", and each field is delimited by a blank. Each reaction must have at least one rate constant, and reversible steps have two rate constants. The numerical value of each constant is preceeded by "/b", and the forward reaction constant is mentioned first. Initial conditions may be specified by preceding the name of the reactant with the initial value followed by a blank. If no initial concentration is stipulated, it is assumed to be zero. The name of a chemical reactant can be up to ten characters long, it must begin with a letter, and it cannot contain embedded blanks or commas. Number fields indicating the numerical values of rate constants and initial conditions can be any legal FORTRAN fixed- or floating-point number up to twenty characters long. All reactant names, initial conditions, rate constants, plus or equal signs, commas, and slashes are followed by a blank.

At the end of the KTL list, control is passed to the Solution Control Language (SCL) interpreter. The purpose of the SCL is to vary rate constants, to set parameters that control solution editing, and to set up series of simulations. Table 4.2 contains a list of SCL commands; the minimum command in any program is STOP. The use of this command causes compilation: The reactions are listed, the total number of reactions and reactants calculated, and syntax is checked. SCL statements can also be used to vary initial conditions and rate constants. If kinetic statements follow an ADD command, the current concentrations are varied. The CONC command allows the concentration of a reactant specified in the NAME field to be changed to the value on the VALUE field. The FLUX statement will set the flux of the reaction in the number field to the value in the VALUE field. LIST statements produce a listing of the

TABLE 4.2. SCL commands used by the University College London Enzyme Simulator.

Command	Operand(s)	Operation
ADD	None	Start new run with current values of concentrations and rate constants.
BEGIN	None	Start new run with initial conditions.
CONC	NAME, VALUE	Set initial and current concentration of variable NAME to VALUE.
*DELTA	VALUE	Set time interval to VALUE. Total time = DELTA. (NSTEP)
ERROR	VALUE	Set relative accuracy of solution to VALUE (relative to the value set by MAX).
FLUX	NUMBER, VALUE	Set current flux of reaction NUMBER to VALUE.
GRAPH	NAME, VALUE, POINT	Plot reactant NAME on scale VALUE displaying character POINT.
HOLD	None	Hold K1 equal to K2...
INPUT	NUMBER	Select input NUMBER.
K	NUMBER, VALUE	Set rate constant of reaction NUMBER to VALUE.
KMAX	None	Set maximum value for optimized constant.
KMIN	None	Set minimum value for optimized constant.
LIST	None	Print current values of concentration, reaction rates, and fluxes.
LIST	NAME	Print current concentration of variable NAME.
MAX	VALUE	NAME1 NAME2 ... Set the scale factor of NAME1 to VALUE.
NSTEP	NUMBER	Set number of integration steps.
PRINT	NUMBER	Select output NUMBER.
RESUME	None	Continue solution.
STOP	None	There is nothing further to be done.
SCALE	None	Scale all data points to current maximum.
TIME	None	NAME1 NAME2 ... provides template for reading experimental data.

current values of all the components. A GRAPH statement plots time versus the reactant in the NAME field scaled by a value in the VALUE field and represented by the symbol mentioned in the POINT field. The time coordinate for GRAPH statements is controlled by the DELTA and NSTEP commands. The DELTA command defines the interval plotted, and the NSTEP command stipulates the fraction of the total integration time to be displayed. A VARY statement causes entry into the SCL interpreter at the time in the VALUE field.

Table 4.3 is a list of the contents of eleven punched cards that were used as input to the University College London Enzyme Simulator. The first card contains a "*" in column 1 followed by the comment "   ONE ENZYME SYSTEM". The next two cards contain the mechanism to be simulated:

$$E + S \underset{k_2}{\overset{k_1}{\rightleftharpoons}} ES \xrightarrow{k_3} E + P \qquad (4.35)$$

Card 2 contains the first elementary reaction, E + S = ES. Columns 1 through 18 stipulate that a reactant called "E" (initially present at concentration 1.0) reacts with a substance called "S" (initially present at concentration 8.0) to form a substance called

TABLE 4.3. Input to simulate equations (4.35) on the University College London Enzyme Simulator.

```
1234567890123456789012345678901234567890
*    ØNE ENZYME SYSTEM ,
1.0 E + 8.0 S = ES   / 1.0 / 1.0 ,
ES = E + P           / 1.0 ,
** ,
DELTA 1.0 ,
NSTEP 20 ,
GRAPH ES 1.0 C ,
GRAPH S 8.0 S ,
GRAPH P 8.0 P ,
BEGIN ,
STØP  ,
```

"ES" (initially absent). Columns 39-49 define the numerical values for the rate constants for the forward reaction (k_1 = 1.0) and the reverse reaction (k_2 = 1.0). Card 3 is the second chemical reaction, ES = E + P. Columns 1-10 stipulate that the reactant called "ES" forms the reactant called "E" and a new substance called "P" (initially absent). Columns 39-43 define k_3 = 1.0 for the rate constant for the forward reaction; since a numerical value for the second rate constant is not given, the constant for the reverse reaction is taken to be zero. Card 4 is the "**b," used to terminate the KTL. Card 5 is the first SCL interpreter command: the DELTA 1.0 command sets the time interval to 1.0. Card 6 sets the number of integration steps to 20; the total integration time plotted is thus DELTA x NSTEP = 20.0 time units. Card 7 causes the substance called "ES" to be plotted as the symbol "C" on a scale with a maximum value of 1.0. Cards 8 and 9 cause S and P to be plotted as "S" and "P" on a scale with a maximum value of 8.0. Card 10 causes the command string of the SCL to be executed, and card 11 terminates the program. All cards in Table 4.3 are terminated with a blank-comma.

Table 4.4 is the output from the University College London Enzyme Simulator when the input is the list of commands in Table 4.3. The comment "   ONE ENZYME SYSTEM" is printed on the first line. The second line contains the reaction number for the first step followed by the rate-constant subscripts for the forward (1) and reverse (2) reactions, the elementary chemical reaction, the initial conditions, and the numerical values for the individual rate constants. The third line contains the reaction number for the second step, the rate constant subscript for the forward step (3), the chemical reaction, and the numerical value for the forward reaction rate constant. The program automatically generates the fourth line to indicate the end of the set of elementary chemical reactions. The fifth line is the total number of chemical steps in the mechanism, the sixth line indicates the total number of forward and reverse reactions, the seventh line gives the total number of reactants, and the eighth line confirms that there are no syntax errors. The next 21 lines are the display of time versus ES, S, and P generated by the DELTA, NSTEP, and GRAPH commands. The user execution time for the simulation in Table 4.4 was 3.7 sec when the calculations were carried out on a Sigma-V computer.

Tables 4.5 and 4.6 are the input and output for the simulation of the 2S, 2P-random mechanism:

$$X_0 + S_0 \underset{k_2}{\overset{k_1}{\rightleftharpoons}} X_1$$

$$X_0 + S_1 \underset{k_4}{\overset{k_3}{\rightleftharpoons}} X_3$$

$$X_2 \underset{k_6}{\overset{k_5}{\rightleftharpoons}} X_3 + S_0$$

$$X_2 \underset{k_8}{\overset{k_7}{\rightleftharpoons}} X_1 + S_1$$

$$X_2 \underset{k_{10}}{\overset{k_9}{\rightleftharpoons}} X_4$$

$$X_4 \underset{k_{12}}{\overset{k_{11}}{\rightleftharpoons}} X_5 + P_0$$

$$X_5 \underset{k_{14}}{\overset{k_{13}}{\rightleftharpoons}} X_0 + P_1$$

$$P_0 + X_0 \underset{k_{16}}{\overset{k_{15}}{\rightleftharpoons}} X_6$$

$$P_1 + X_6 \underset{k_{18}}{\overset{k_{17}}{\rightleftharpoons}} X_4 \qquad (4.36)$$

The first two input lines are the comments printed on the first two output lines. The next nine input lines contain the reaction num-

TABLE 4.4. Output for Eqs. (4.35) from the University College London Enzyme Simulator.

```
     1     1     2 1•0 E + 8•0 S = ES                           / 1•0 / 1•0 ,
     2     3         ES = E + P      .                          / 1•0 ,
                 ******END EQUATIØN LIST
EQUATIØNS  =       2
REACTIØNS  =       3
VARIABLES  =       4
ERRØRS =       0
    •0000 •
   1•0000 •   P                                 C S
   2•0000 •        P                          SC
   3•0000 •            P                 S    C
   4•0000 •                 P        S      C
   5•0000 •                     PS        C
   6•0000 •                  S      P    C
   7•0000 •               S           C P
   8•0000 •            S            C      P
   9•0000 •         S            C            P
  10•0000 •       S           C                  P
  11•0000 •     S          C                       P
  12•0000 •   S         C                            P
  13•0000 •  S       C                                 P
  14•0000 • S     C                                     P
  15•0000 •S    C                                        P
  16•0000 •S  C                                           P
  17•0000 •  C                                            P
  18•0000 • C
  19•0000 •C
  20•0000 •C
```

TABLE 4.5. Input to simulate Eqs. (4.36) on the University College London Enzyme Simulator.

```
*1-ENZYME, 2-SUBSTRATE, 2-PRØDUCT SYSTEM ,
*RANDØM SEQUENCE ,
1.0 XO + 1.0 SO = X1                  / 1.0 / 1.0 ,
X1 + 5.0 S1 = X2                      / 0.1 / 0.1 ,
X2 = X3 + SO                          / 1.0 / 1.0 ,
X3 = XO + S1                          / 10.0 / 10.0 ,
X2 = X4                               / 0.02 / 0.02 ,
X4 = X5 + PO                          / 10.0 / 10.0 ,
X5 = XO + P1                          / 1.0 / 1.0 ,
X4 = X6 + P1                          / 0.1 / 0.1 ,
X6 = XO + PO                          / 1.0 / 1.0 ,
** ,
DELTA 0.1 ,
NSTEP 25 ,
GRAPH SO 1.0 S ,
GRAPH X2 1.0 C ,
BEGIN ,
STØP ,
```

TABLE 4.6. Output for Eqs. (4.36) from the University College London Enzyme Simulator.

```
                   1-E, 2-S, 2-P SYSTEM ,

                   RANDOM SEQUENCE ,
   1     1      2 1•0 X0 + 1•0 S0 = X1              / 1•0 / 1•0 ,
   2     3      4 X1 + 5•0 S1 = X2                  / 0•1 / 0•1 ,
   3     5      6 X2 = X3 + S0                      / 1•0 / 1•0 ,
   4     7      8 X3 = X0 + S1                      / 10•0 / 10•0 ,
   5     9     10 X2 = X4                           / 0•02 / 0•02 ,
   6    11     12 X4 = X5 + P0                      / 10•0 / 10•0 ,
   7    13     14 X5 = X0 + P1                      / 1•0 / 1•0 ,
   8    15     16 X4 = X6 + P1                      / 0•1 / 0•1 ,
   9    17     18 X6 = X0 + P0                      / 1•0 / 1•0 ,
             ******END EQUATION LIST
EQUATIONS  =     9
REACTIONS  =    18
VARIABLES  =    11
ERRORS =      0
 •0000 •
 •1000 •  C                                          S
 •2000 •     C                                    S
 •3000 •       C                               S
 •4000 •         C                           S
 •5000 •          C                         S
 •6000 •           C                       S
 •7000 •            C                     S
 •8000 •            C                    S
 •9000 •             C                   S
1•0000 •             C                  S
1•1000 •              C                 S
1•2000 •              C                S
1•3000 •              C                S
1•4000 •              C                S
1•5000 •              C                S
1•6000 •              C                S
1•7000 •              C                S
1•8000 •              C                S
1•9000 •              C               S
2•0000 •              C               S
2•1000 •              C               S
2•2000 •               C              S
2•3000 •               C              S
2•4000 •               C              S
2•5000 •               C              S
```

bers and rate-constant subscripts for each of the nine reversible chemical reactions (eighteen chemical steps), the initial conditions for the eleven variables, and the values for the eighteen rate constants: $E(0) = 1.0$, $S_0(0) = 1.0$, $S_1(0) = 5.0$, $X_1(0) = X_2(0) = X_3(0) = X_4(0) = X_5(0) = X_6(0) = P_0(0) = P_1(0) = 0$, $k_1 = 1.0$, $k_2 = 1.0$, $k_3 = 0.1$, $k_4 = 0.1$, $k_5 = 1.0$, $k_6 = 1.0$, $k_7 = 10.0$, $k_8 = = 10.0$, $k_9 = 0.02$, $k_{10} = 0.02$, $k_{11} = 10.0$, $k_{12} = 10.0$, $k_{13} = 1.0$,

$k_{14} = 1.0$, $k_{15} = 0.1$, $k_{16} = 0.1$, $k_{17} = 1.0$, nad $k_{18} = 1.0$. These equations, initial conditions and rate constants are output in lines 3-11 in Table 4.6. Line 12 signals the end of the equation list, line 13 indicates the total number of chemical steps, line 14 the total number of forward or reverse reactions, line 15 the total number or reactants, and line 16 confirms that there are no syntax errors. The last 26 lines are the display of time versus S_0 (plotted as "S") and X_2 (plotted as "C"). These plots are generated by the DELTA, NSTEP, and GRAPH commands in Table 4.5. The user execution time for the simulation in Table 4.6 was 8.8 sec when the calculations were carried out on a Sigma-V computer.

CHAPTER 5
ANALYSIS OF ENZYME KINETIC DATA

5.1 THE TRADITIONAL APPROACH

Enzyme kinetic data usually consist of the concentration of a product or substrate measured as a function of time. The usual procedure is to assume that during the early stages of the overall reaction there is a quasi-steady state for the enzyme intermediates. Since data obtained after the overall reaction is a few percent complete are often viewed with suspicion, they are commonly discarded! The time-honored method for the "analysis" of the data is to place something with a straightedge "tangent" to the "earliest" possible portion of $P(t)$ or $S(t)$. This visually identified tangent is assumed to be the "initial, quasi-steady-state rate" $v(0)$. The experiment is repeated for different initial substrate concentrations $[S(0)]$, and $1/v(0)$ is plotted versus $1/S(0)$. The result is called a Lineweaver-Burke plot (notwithstanding the fact that it was first suggested for this purpose by Woolf[14]). If this plot seems linear, a straight line is drawn through the points; the reciprocal of the $1/v(0)$ axis intercept is called the maximum velocity V_m, and the ratio of the slope to this intercept is called the Michaelis constant K_m. Usually, neither the *real* data from which the rates were estimated nor the details of how these estimations were carried out are published. It is not surprising when such results cannot be reproduced in other laboratories.

The methods just described have very little merit, and improvements have been suggested from time to time. Walker and Schmidt[15] proposed a method of plotting and using a greater portion of the original data, and Fleisher[16] proposed an improvement that permitted an objective estimate of v(0) from P(t) or S(t). As usual, either of these methods is applicable only over short time intervals when the quasi-steady-state approximation is valid, and not over the longer time intervals suggested by Walker, Schmidt, and Fleischer. Hofstee[17] pointed out that the reciprocal method of plotting initial rates and substrate concentrations could be misleading, and he emphasized the advantages of v(0) versus v(0)/S(0) plots to detect model errors. Hofstee emphasized that plots of 1/v(0) versus 1/S(0) tend to obscure experimental inferences that the classical enzyme model is incorrect as well as various kinds of systematic or random errors in the plotted quantities. Notwithstanding these criticisms,[18] 97.3% of the "linear plots" in the 1971 issue of the *Journal of Biological Chemistry* were reciprocal plots; only 1.6% were the demonstrably superior[19] v(0) versus v(0)/S(0) plots, and the remaining 1.1% were the grossly inferior S(0)/v(0) versus S(0) type plot. Since 20.2% of all papers in the 1971 issue of the *Journal of Biological Chemistry* uses one of these linear forms, and since each of these papers contained an average of 10.1 linear plots, the relationship between initial, quasi-steady-state rates and initial substrate concentrations must be vital to the results published in this prestigeous journal. It is not clear, however, why the results are presented in the 1/v(0) versus 1/S(0) form.

5.2 CALCULATIONS AND GRAPHICAL METHODS USEFUL IN THE ANALYSIS OF ESTIMATED INITIAL QUASI-STEADY-STATE RATES

The two simplest mathematical models relating initial quasi-steady-state rates and one of the initial conditions are

$$v(0) = \frac{\alpha_1 x}{1 + \beta_1 x} \tag{5.1}$$

and

$$v(0) = \frac{\alpha_1 x + \alpha_2 x^2}{1 + \beta_1 x + \beta_2 x^2} \tag{5.2}$$

where x is the initial concentration of a substrate or modifier (i.e., an "inhibitor" or "stimulator"), and α and β are constant coefficients that depend on the other initial conditions and the individual rate constants in the chemical model. More complex models involving higher-order powers of x are clearly possible [See Eq. (1.39)], but Eqs. (5.1) and (5.2) are the two *simplest* possible models. In what follows we refer to Eq. (5.1) as the "degree-one model" (because it is described by a rational function of degree one), and to Eq. (5.2) as the "degree-two model." For quasi-steady-state models the degree-one model results from mechanisms such as those involving one binding site on the enzyme (single-substrate mechanisms, mechanisms of competitive inhibition, etc.), multiple, noninteracting binding sites, or ordered sequences of substrate addition. The degree-two model, on the other hand, results from mechanisms involving interacting binding sites on the enzyme (not competitive[20] inhibition, not competitive stimulation, random- or preferred-ordered mechanisms, etc.).

The minimum aim of any scientific investigation should be to distinguish between the two simplest mathematical models. In the case of quasi-steady-state kinetic studies with enzyme-catalyzed reactions, the goal is to decide if the data fit Eq. (5.1), of if Eq. (5.2) (or a rational function of higher degree) must be used. The decision is easy to make if the experimental data contain a feature such as an extremum or an inflection: Mechanisms described by Eq. (5.1) cannot have an extremem or an inflection in plots of v(0) versus S(0), but mechanisms described by Eq. (5.2) have an extremum if

$$\frac{\partial v(0)}{\partial x} = \frac{\alpha_1 + 2\alpha_2 x + (\alpha_2\beta_1 - \alpha_1\beta_2)x^2}{(1 + \beta_1 x + \beta_2 x^2)^2} = 0 \tag{5.3}$$

for positive, real x. In other words, if $\alpha_1\beta_2 > \alpha_2\beta_1$, the degree-two model possesses a feature (an extremum) that cannot be present in the degree-one model. Furthermore, if

$$\frac{\partial^2 v(0)}{\partial x^2} = \frac{\alpha_2 - \alpha_1\beta_1 - 3\alpha_1\beta_2 x - 3\alpha_2\beta_2 x^2 - \beta_2(\alpha_2\beta_1 - \alpha_1\beta_2)x^3}{(1 + \beta_1 x + \beta_2 x^2)^3} = 0 \tag{5.4}$$

the degree-two model possesses an inflection point in plots of v(0) versus x. If $\alpha_1\beta_2 > \alpha_2\beta_1$, there must be an inflection between the extremum and asymptotic values of v(0). If $\alpha_1\beta_2 > \alpha_2\beta_1$ *and* $\alpha_2 > \alpha_1\beta_1$, there is also an inflection between zero and the extremum value of v(0). Finally, if $\alpha_1\beta_2 < \alpha_2\beta_1$, there is no extremum, but if $\alpha_2 > \alpha_1\beta_1$, there is an inflection point. In all these cases the degree-two model possesses a feature that distinguishes it from the degree-one model. Therefore, if experimental data indicate that plots of v(0) versus x possess an extremum or an inflection, the degree-one model can be eliminated from consideration.

It is clear from Eqs. (5.1) and (5.2) that the degree-two model converges to the degree-one model as α_2 and β_2 approach zero. This means that for small α_2 and β_2 it will be difficult or impossible to distinguish between these two models. However, in such cases the numerical values for α_1 and β_1 calculated for each model will be similar.

Even if α_2 and β_2 are relatively large, it may not always be possible to distinguish between a hyperbola generated from Eq. (5.2) when $\alpha_1\beta_2 < \alpha_2\beta_1$ and $\alpha_2 < \alpha_1\beta_1$ and a hyperbola generated for a different set of α_1 and β_1 from Eq. (5.1). Statistical fits of real data to Eqs. (5.1) and (5.2) may lead to different sets of parameters, each of which possesses a minimum sum of squared deviations. When the minimum sums produced by the fit to each model are sufficiently similar that they cannot be experimentally distinguished, the two models become indistinguishable. In this case, however, the numerical values of α_1 and β_1 calculated for each model will be different.

The usual method of distinguishing between the degree-one model and higher-degree models such as Eq. (5.2) is to plot the experimental quantities in a "biased" form that should be linear for the degree-one model, but nonlinear for the degree-two model. In such cases it should be possible to detect a specific concave trend if the biased plot is not linear. The use of graphical procedures of this type depends upon finding a form that is linear for one model, but markedly nonlinear for the other model. It is also extremely important to collect data in a manner that is free of systemic error (since systemic error will inevitably show up as a specific concave trend even in the model for which the biased plot should be linear), and it is desirable to collect data with a reasonably high precision.

Equation (5.1) can be rearranged to the linear forms

$$\frac{1}{v(0)} = \frac{\beta_1}{\alpha_1} + \frac{1}{\alpha_1}\frac{1}{x} \tag{5.5}$$

$$\frac{x}{v(0)} = \frac{1}{\alpha_1} + \frac{\beta_1}{\alpha_1} x \tag{5.6}$$

$$v(0) = \frac{\alpha_1}{\beta_1} - \frac{1}{\beta_1}\frac{v(0)}{x} \tag{5.7}$$

Equation (5.5) illustrates that for the degree-one model, plots of $1/v(0)$ versus $1/x$ should be linear; the $1/v(0)$ axis intercept is β_1/α_1, and the slope is $1/\alpha_1$. Equation (5.6) illustrates that for this model, plots of $x/v(0)$ versus x should also be linear; the $x/v(0)$ axis intercept is $1/\alpha_1$, and the slope is β_1/α_1. And Eq. (5.7) illustrates that plots of $v(0)$ versus $v(0)/x$ for the degree-one model should be linear; the v(0) axis intercept is α_1/β_1, and the slope is $-1/\beta_1$.

If one attempts to rearrange Eq. (5.2) into any of the forms that are linear for Eq. (5.1), one does not obtain linear relationships. The slope of plots of 1/v(0) versus 1/x for the degree-two model is

$$\frac{\partial[1/v(0)]}{\partial(1/x)} = \frac{\alpha_1 + 2\alpha_2 x + (\alpha_2\beta_1 - \alpha_1\beta_2)x^2}{(\alpha_1 + \alpha_2 x)^2} \tag{5.8}$$

the slope of plots of x/v(0) versus x is

$$\frac{\partial[x/v(0)]}{\partial x} = \frac{\alpha_1\beta_1 - \alpha_2 + 2\alpha_1\beta_2 x + \alpha_2\beta_2 x^2}{(\alpha_1 + \alpha_2 x)^2} \tag{5.9}$$

and the slope of plots of v(0) versus v(0)/x is

$$\frac{\partial v(0)}{\partial[v(0)/x]} = -\frac{\alpha_1 + 2\alpha_2 x + (\alpha_2\beta_1 - \alpha_1\beta_2)x^2}{\alpha_1\beta_1 - \alpha_2 + 2\alpha_1\beta_2 x + \alpha_2\beta_2 x^2} \tag{5.10}$$

Since $\alpha_2\beta_1 > \alpha_1\beta_2$ and $\alpha_1\beta_1 > \alpha_2$, the slopes of plots of 1/v(0) versus 1/x or x/v(0) versus x are always positive, and the slope of plots of v(0) versus v(0)/x is always negative. In each case, however, the slope varies with x.

The curvature in plots of 1/v(0) versus 1/x, x/v(0) versus x, or v(0) versus v(0)/x for the degree-one model is zero, but for the degree-two model

$$\frac{\partial^2[1/v(0)]}{\partial(1/x)^2} = \frac{2x^3[\alpha_2^2 + \alpha_1(\alpha_1\beta_2 - \alpha_2\beta_1)]}{(\alpha_1 + \alpha_2 x)^3} \tag{5.11}$$

$$\frac{\partial^2[x/v(0)]}{\partial x^2} = \frac{2[\alpha_2^2 + \alpha_1(\alpha_1\beta_2 - \alpha_2\beta_1)]}{(\alpha_1 + \alpha_2 x)^3} \tag{5.12}$$

and

$$\frac{\partial^2 v(0)}{\partial[v(0)/x]^2} = -\frac{2[\alpha_2^2 + \alpha_1(\alpha_1\beta_2 - \alpha_2\beta_1)](1 + \beta_1 x + \beta_2 x^2)^3}{(\alpha_1\beta_1 - \alpha_2 + 2\alpha_1\beta_2 x + \alpha_2\beta_2 x^2)^3} \tag{5.13}$$

In terms of the scaled quantities

$$\sigma \equiv \alpha_1 x \quad \text{and} \quad V \equiv v(0)\ \beta_2/\alpha_2 \tag{5.14}$$

the curvature equations are

$$C_1 \equiv \frac{\partial^2(1/V)}{\partial(1/\sigma)^2} = \frac{2a_2\sigma^3[a_2 + a_1(a_1 - 1)]}{(a_1 + a_2\sigma)^3} \tag{5.15}$$

$$C_2 \equiv \frac{\partial^2(\sigma/V)}{\partial\sigma^2} = \frac{2a_2[a_2 + a_1(a_1 - 1)]}{(a_1 + a_2\sigma)^3} \tag{5.16}$$

and

$$C_3 \equiv \frac{\partial^2 V}{\partial(V/\sigma)^2} = -\frac{2a_2[a_2 + a_1(a_1 - 1)](1 + \sigma + a_2\sigma^2)^3}{(a_1 - a_2 + 2a_1a_2\sigma + a_2^2\sigma^2)^3} \tag{5.17}$$

where

$$a_1 = \frac{\alpha_1\beta_2}{\alpha_2\beta_1} < 1 \quad \text{and} \quad a_2 = \frac{\beta_2}{\beta_1^2} < a_1 \tag{5.18}$$

The absolute value of the ratio of the curvature in plots of $1/V$ versus $1/\sigma$ to the curvature in plots of V versus V/σ is

$$\left|\frac{C_1}{C_3}\right| = \frac{\sigma^3(a_1 - a_2 + 2a_1a_2\sigma + a_2^2\sigma^2)^3}{(1 + \sigma + a_2\sigma^2)^3(a_1 + a_2\sigma)^3} \tag{5.19}$$

Taking the cube root and multiplying out the terms, we find

$$\left|\frac{C_1}{C_3}\right|^{1/3} = \frac{(a_1 - a_2)\sigma + 2a_1a_2\sigma^2 + a_2^2\sigma^3}{a_1 + (a_1 + a_2)\sigma + a_2(a_1 + 1)\sigma^2 + a_2^2\sigma^3} < 1 \tag{5.20}$$

where $a_1 + a_2 > a_1 - a_2$ because $a_2 > 0$, and $a_1 + 1 > 2a_1$ because $a_1 < 1$. Inequality (5.20) illustrates that for all $0 \leq \sigma < \infty$ in the degree-two model, plots of V versus V/σ possess more curvature than do plots of $1/V$ versus $1/\sigma$ constructed from the same data.[19]

The absolute value of the ratio of the curvature in plots of σ/V versus σ to the curvature in plots of V versus V/σ is

$$\left|\frac{c_2}{c_3}\right| = \frac{(a_1 - a_2 + 2a_1a_2\sigma + a_2{}^2\sigma^2)^3}{(1 + \sigma + a_2\sigma^2)^3(a_1 + a_2\sigma)^3} \tag{5.21}$$

In this case we find that

$$\left|\frac{c_2}{c_3}\right|^{1/3} = \frac{a_1 - a_2 + 2a_1a_2\sigma + a_2{}^2\sigma^2}{a_1 + (a_1 + a_2)\sigma + a_2(a_1 + 1)\sigma^2 + a_2{}^2\sigma^3} < 1 \tag{5.22}$$

where $a_1 > a_1 - a_2$ because $a_2 > 0$, and $a_1 + a_2 > 2a_1a_2$ and $a_1 + 1 > a_2$ because $1 > a_1 > a_2$. Inequality (5.22) illustrates that for all $0 < \sigma < \infty$ in the degree-two model, plots of V versus V/σ possess more curvature than do plots of σ/V versus σ constructed from the same data.[19]

Table 5.1 contains a list of the minimum and maximum possible slopes and curvatures for the degree-two model when V and σ are plotted in each of the forms that would be linear for the degree-one model. Inequalities (5.20) and (5.22) and Table 5.1 illustrate that the V versus V/σ [or the v(0) versus v(0)/x] type plot is the best type to use to detect deviations from the degree-one model.

Figure 5.1 illustrates the case when it is not possible to distinguish between hyperbolas generated from Eqs. (5.1) and (5.2). The actual model is Eq. (5.2), when $\alpha_1 = 3 \times 10^3$, $\beta_1 = 10^4$, $\alpha_2 = 10^7$, and $\beta_2 = 2 \times 10^7$. In Fig. 5.1 (a) the initial rates generated with typical experimental error for this model are plotted versus x, the initial substrate concentration. The solid line in this figure is the plot of v(0) versus x generated directly from Eq. (5.2) (without error). Note that the data look very much like what one would expect for the degree-one model.

If the data in Figure 5.1 (a) fits the degree-one model, we can obtain estimates for α_1 and β_1 from a weighted fit to the linear forms of Eq. (5.1) by minimizing a weighted sum of the squares of the deviations: $\Sigma\, w[v(0) - \bar{v}(0)]^2$. For Eq. (5.6), the approximate weighting factor is

$$w \simeq \frac{v(0)^4}{x^2} \tag{5.23}$$

TABLE 5.1. Maximum curvatures and slopes for transformed plots of the variables in Eq. (5.2).[a]

Type Plot	Slope		Curvature	
	Low σ	High σ	Low σ	High σ
$\frac{1}{V}$ vs. $\frac{1}{\sigma}$	$\frac{1}{a_1}$	$\frac{1 - a_1}{a_2}$	$\longrightarrow 0$ Minimum	$\frac{2[a_2 + a_1(a_1 - 1)]}{a_2^2}$ Maximum
$\frac{\sigma}{V}$ vs. σ	$\frac{a_1 - a_2}{a_1}$	1	$\frac{2a_2[a_2 + a_1(a_1 - 1)]}{a_1^3}$ Maximum	$\longrightarrow 0$ Minimum
V vs. $\frac{V}{\sigma}$	$\frac{a_1}{a_2 - a_1}$	$\frac{a_1 - 1}{a_2}$	$\frac{2a_2[a_2 + a_1(a_1 - 1)]}{(a_2 - a_1)^3}$ Minimum	$\frac{-2[a_2 + a_1(a_1 - 1)]}{a_2^2}$ Maximum

[a] $\sigma = \alpha_1 x$, $V = v(0)\beta_2/\alpha_2$, $a_1 = \alpha_1\beta_2/\alpha_2\beta_1$, $a_2 = \beta_2/\beta_1^2$.

The weighted means for x/v(0) and x are

$$\frac{\bar{x}}{v(0)} = \frac{\Sigma\, v(0)^3/x}{\Sigma\, v(0)^4/x^2} \tag{5.24}$$

$$\bar{x} = \frac{\Sigma\, v(0)^4/x}{\Sigma\, v(0)^4/x^2} \tag{5.25}$$

$$\frac{\beta_1}{\alpha_1} = \frac{\Sigma\, v(0)^3 - \dfrac{[\Sigma\, v(0)^4/x][\Sigma\, v(0)^3/x]}{\Sigma\, v(0)^4/x^2}}{\Sigma\, v(0)^4 - \dfrac{[\Sigma\, v(0)^4/x]^2}{\Sigma\, v(0)^4/x^2}} \tag{5.26}$$

$$\frac{1}{\alpha_1} = \frac{\bar{x}}{v(0)} - \frac{\beta_1}{\alpha_1}\bar{x} \tag{5.27}$$

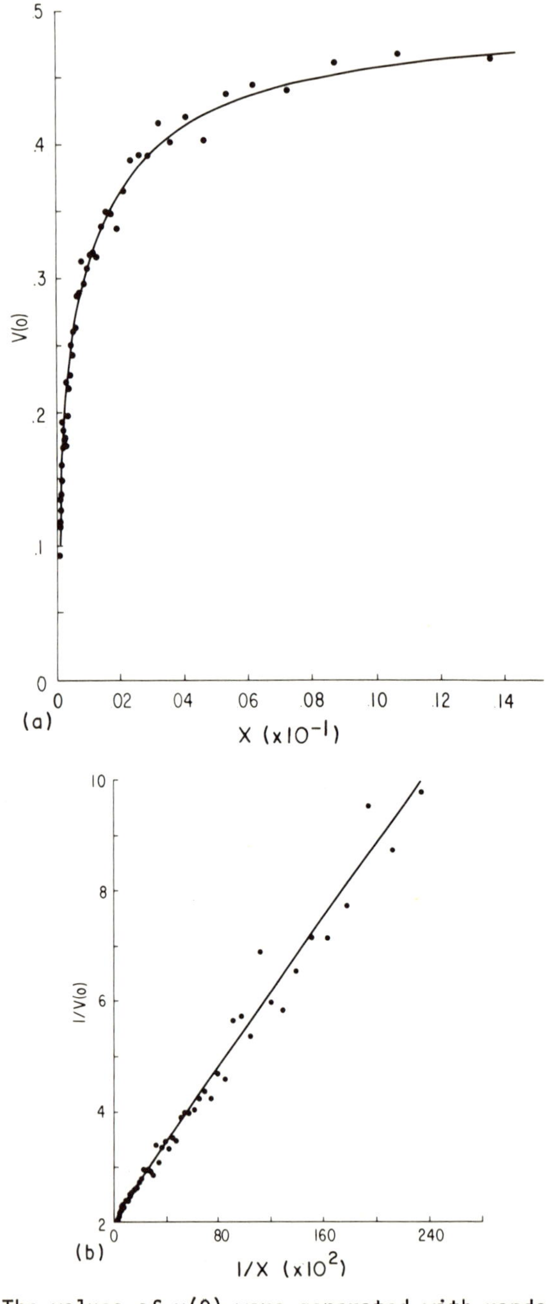

FIG. 5.1. The values of v(0) were generated with random error [10% when v(0) equals one-tenth its asymptotic value, 1% when v(0) equals its asymptote, and proportionate in between] from Eq. (5.2) when $\alpha_1 = 3 \times 10^3$, $\beta_1 = 10^4$, $\alpha_2 = 10^7$, and $\beta_2 = 2 \times 10^7$. The solid lines are the exact function [Eq. (5.2)] with no error.

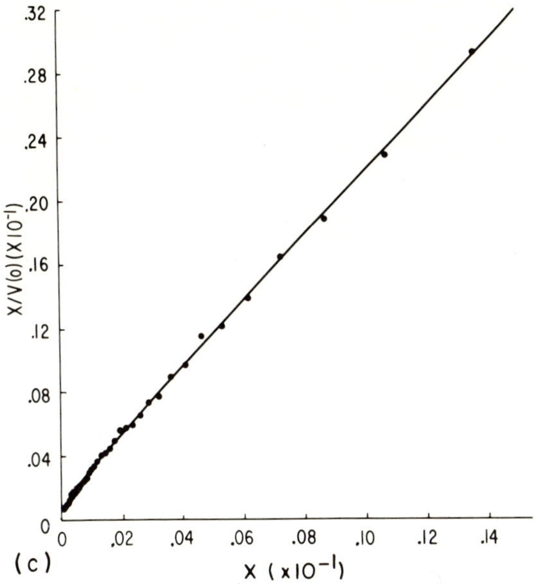
.32
.28
.24
.20
.16
.12
.08
.04
0
X/V(o)(X10^{-1})
0
.02
.04
.06
.08
.10
.12
.14
X (x10^{-1})

(c)

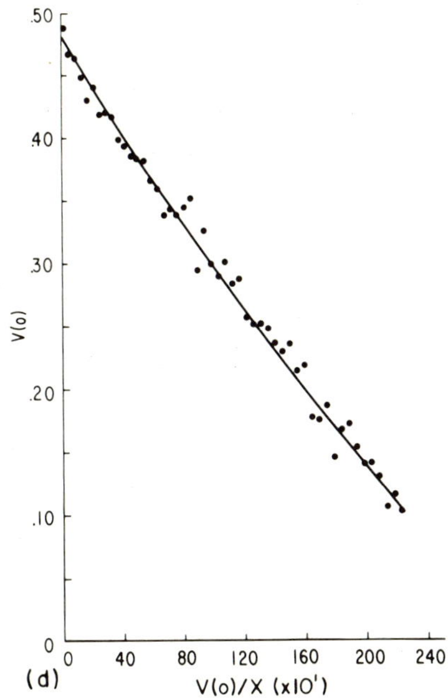
.50
.40
.30
.20
.10
0
V(o)
0
40
80
120
160
200
240
V(o)/X (x10^{1})

(d)

Figures 5.1 (b), (c), and (d) are the three types of plots that should be linear for Eq. (5.1). The plotted quantities are the transformed variables calculated from the plotted quantities in Fig. 5.1 (a), and the solid lines are the values of the transformed variables generated directly from Eq. (5.2). Note that all three plots appear linear: the data still look very much like what one would expect for the degree-one model. However, the value of $\beta_1 = 6.2 \times 10^3$ calculated for the degree-one model from a weighted least-squares fit to the linear forms of Eq. (5.1) is different from the value used to generate the data from Eq. (5.2). Furthermore, if one uses these weighted estimates for α_1 and β_1 ($\alpha_1 = 3 \times 10^3$ and $\beta_1 = 6.2 \times 10^3$) together with $\alpha_2 = 0$ or $\alpha_2 = \alpha_1{}^2$ and $\beta_2 = 0$ or $\beta_2 = \beta_1{}^2$ as the starting values for a direct nonlinear fit to Eq. (5.2), convergence criteria based on a 0.001% improvement in the sum of squares of the deviations will not improve the values of the parameters by more than a percent or two in twenty iterations. In other words, the sums of the squares for the degree-one and -two models are so similar that one cannot distinguish experimentally between them when $\alpha_1 = 3 \times 10^3$ and $\beta_1 = 6.2 \times 10^3$ for the degree-one model, and $\alpha_1 = 3 \times 10^3$, $\beta_1 = 10^4$, $\alpha_2 = 10^7$, and $\beta_2 = 2 \times 10^7$ for the degree-two model. The final values for these parameters calculated from the direct least-squares fit for the data with error in Fig. 5.1 is $\alpha_1 = 3.146948 \times 10^3$, $\beta_1 = 1.032803 \times 10^4$, $\alpha_2 = 9.962223 \times 10^6$, and $\beta_2 = 2.002544 \times 10^7$. The significance level for the distinction between Eqs. (5.1) and (5.2) calculated from the variance ratio for the two models and fifty experimental determinations is in the range of 50%. The significance level is the number of times out of a hundred that the data will lead to the correct model. If one flips a coin to decide between Eq. (5.1) and (5.2), the significance level of the decision would be 50%.

Figure 5.2 illustrates the case when it is relatively easy to distinguish between hyperbolas generated from Eqs. (5.1) and (5.2). The actual model is again Eq. (5.2), but in Fig. 5.2 $\alpha_1 = 2 \times 10^3$, $\beta_1 = 8 \times 10^3$, $\alpha_2 = 2 \times 10^6$, and $\beta_2 = 4 \times 10^6$. The initial rates generated with typical experimental error for this model are plotted versus x in Fig. 5.2 (a). The solid line is the plot of v(0) versus x generated directly from Eq. (5.2) without error. Note that the data in Fig. 5.2 (a) look very much like what one would expect for the degree-one model.

Figures 5.2 (b), (c), and (d) are the three type plots that should be linear for Eq. (5.1). The plotted quantities are the transformed variables calculated from the plotted quantities in Fig. 5.2 (a), and the solid lines are the values of the transformed variables generated directly from Eq. (5.2). In this case the 1/v(0) versus 1/x plot [Fig. 5.2 (b)] is curved, but the curvature is not obvious at lower v(0). The x/v(0) versus x plot [Fig. 5.2 (c)] is also curved, but the curvature is not obvious at all. The v(0) versus v(0)/x plot [Fig. 5.2 (d)] possesses the most obvious curvature of the three types of transformed plots, and it is therefore the best type to use to detect deviations from Eq. (5.1). The significance level for the distinction between Eqs. (5.1) and (5.2) based on fifty experimental determinations is 99.9%, and the significance level based on twelve random determinations of v(0) is 91%. Thus, when $\alpha_1 = 2 \times 10^3$, $\beta_1 = 8 \times 10^3$, $\alpha_2 = 2 \times 10^6$, and $\beta_2 = 4 \times 10^6$, one can use v(0) versus v(0)/x plots or statistical calculations of significance levels to ascertain that the data [Fig. 5.2 (a)] fit Eq. (5.2) much better than they fit Eq. (5.1), but the statistical calculations depend on the availability of what in practical terms is a rather large number of experimental determinations.

Figure 5.3 illustrates a case when it is relatively difficult to distinguish between hyperbolas generated from Eqs. (5.1) and (5.2). The actual model is once again Eq. (5.2), but in Fig. 5.3 $\alpha_1 = 2 \times 10^3$, $\beta_1 = 8 \times 10^3$, $\alpha_2 = 2 \times 10^6$, and $\beta_2 = 5.5 \times 10^6$. The initial rates generated with typical experimental error for this model are plotted versus x in Fig. 5.3 (a). The solid line is the plot of v(0) versus x generated directly from Eq. (5.2) without error.

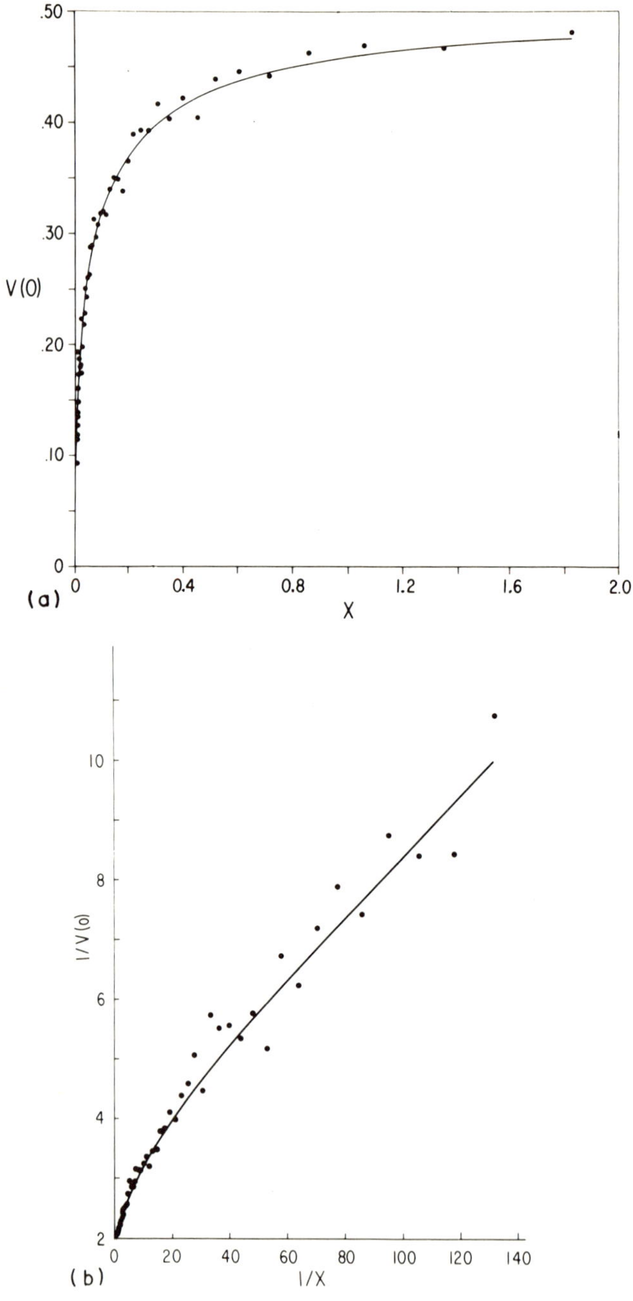

FIG. 5.2. The values of v(0) were generated as in Fig. 5.1, except $\alpha_1 = 2 \times 10^3$, $\beta_1 = 8 \times 10^3$, $\alpha_2 = 2 \times 10^6$, and $\beta_2 = 4 \times 10^6$. The solid lines are the exact function with no error.

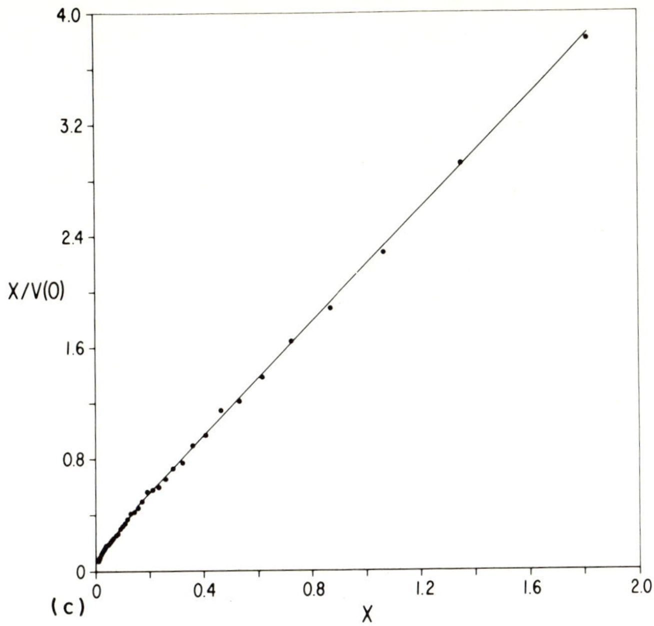
4.0
3.2
2.4
X/V(0)
1.6
0.8
0
0
0.4
0.8
1.2
1.6
2.0
X

(c)

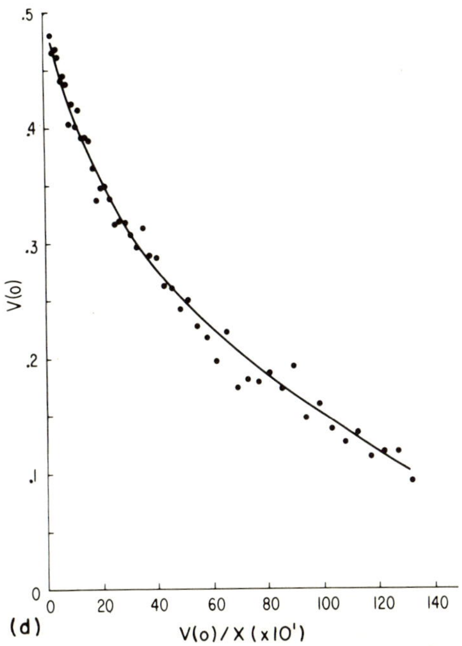
.5
.4
.3
V(o)
.2
.1
0
0
20
40
60
80
100
120
140
V(o)/X (x10')

(d)

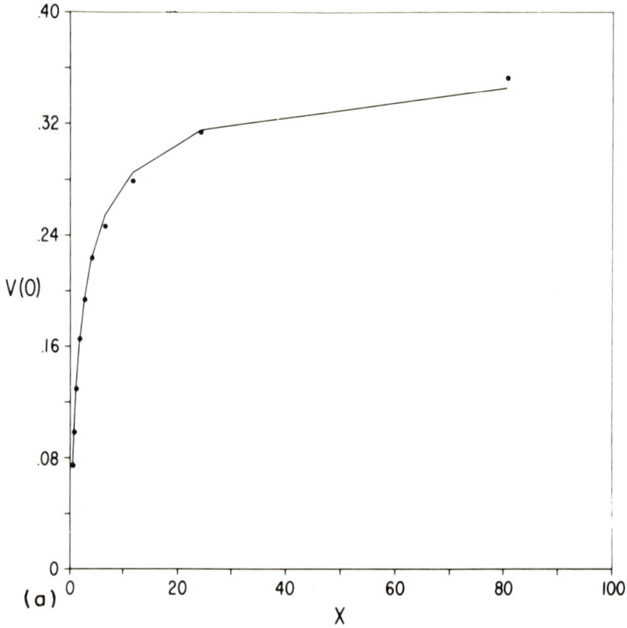

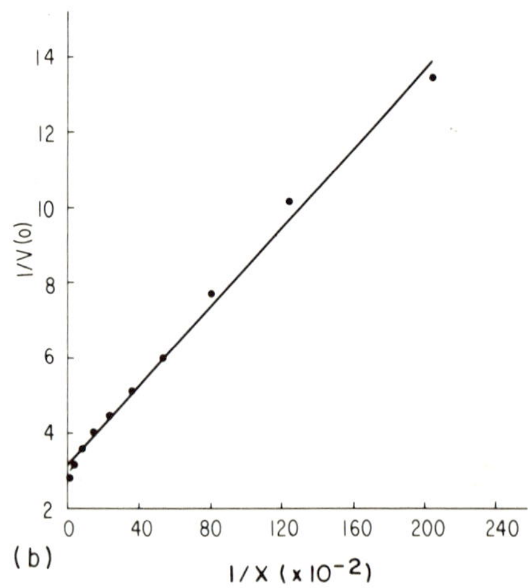

FIG. 5.3. The values of v(0) were generated as in Fig. 5.2, except $\beta_2 = 5.5 \times 10^6$. The solid line in (a) is the exact function with no error, and the solid lines in (b)-(d) are the unweighted regression fits to the linear forms of Eq. (5.1).

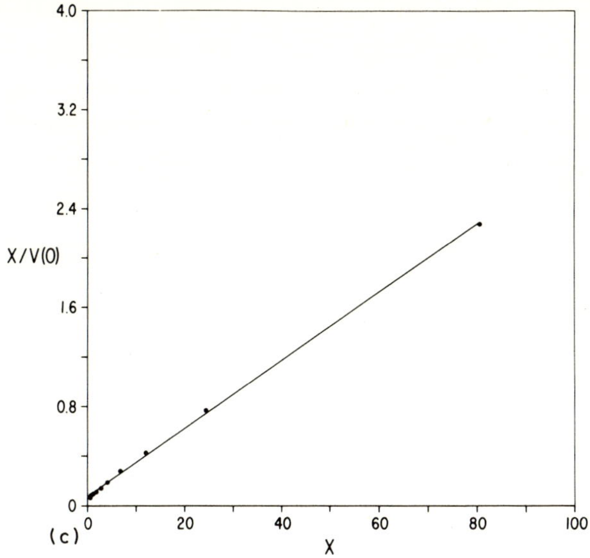
4.0
3.2
2.4
X/V(0)
1.6
0.8
0
0
20
40
60
80
100
X

(c)

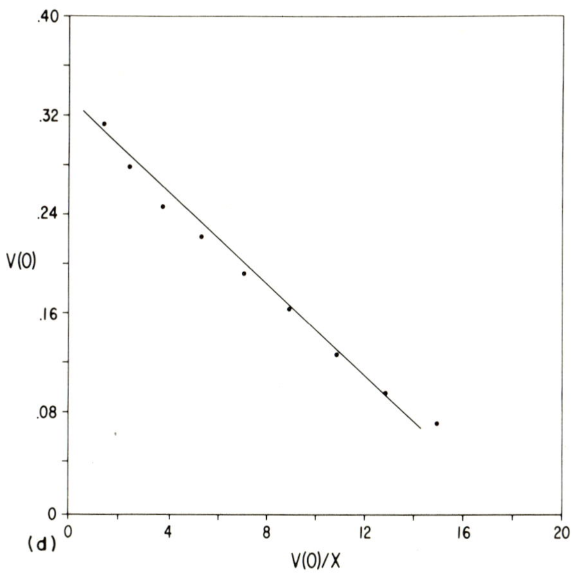
.40
.32
.24
V(0)
.16
.08
0
0
4
8
12
16
20
V(0)/X

(d)

Figures 5.3 (b), (c), and (d) are again the three type plots that should be linear for Eq. (5.1). The solid lines are the unweighted regression fits to the respective linear forms of Eq. (5.1). In this case, only the v(0) versus v(0)/x plot [Fig. 5.3 (d)] possesses obvious curvature. The average significance level for distinguishing the two models on the basis of ten to a hundred determinations appears in Fig. 5.4. These significance levels range from 70% at ten data points up to 95% at a hundred points. This means that to obtain the usual level of significance to distinguish models on the basis of statistical calculations, one would have to determine over a hundred initial rates.

In Fig. 5.5 appears the relationship between σ and the absolute value of the normalized curvature for the solid lines in Fig. 5.1 (b) ($1/V$ versus $1/\sigma$), 5.1 (c) (σ/V versus σ), and 5.1 (d) (V versus V/σ). As anticipated from inequalities (5.20) and (5.22), the curvature in plots of V versus V/σ is always larger than the curvature in plots of $1/V$ versus $1/\sigma$ or σ/V versus σ constructed from the same data.

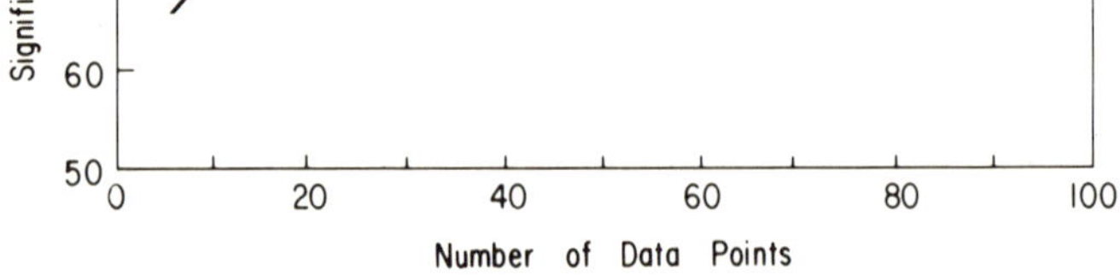

FIG. 5.4. The average significance levels were calculated from 25 sets of v(0) and x for each number of data points. The individual significance levels were calculated from the ratio of the variances obtained by fitting the points to Eqs. (5.1) and (5.2). Each set of v(0) and x used to construct this figure was generated independently as described in Fig. 5.3.

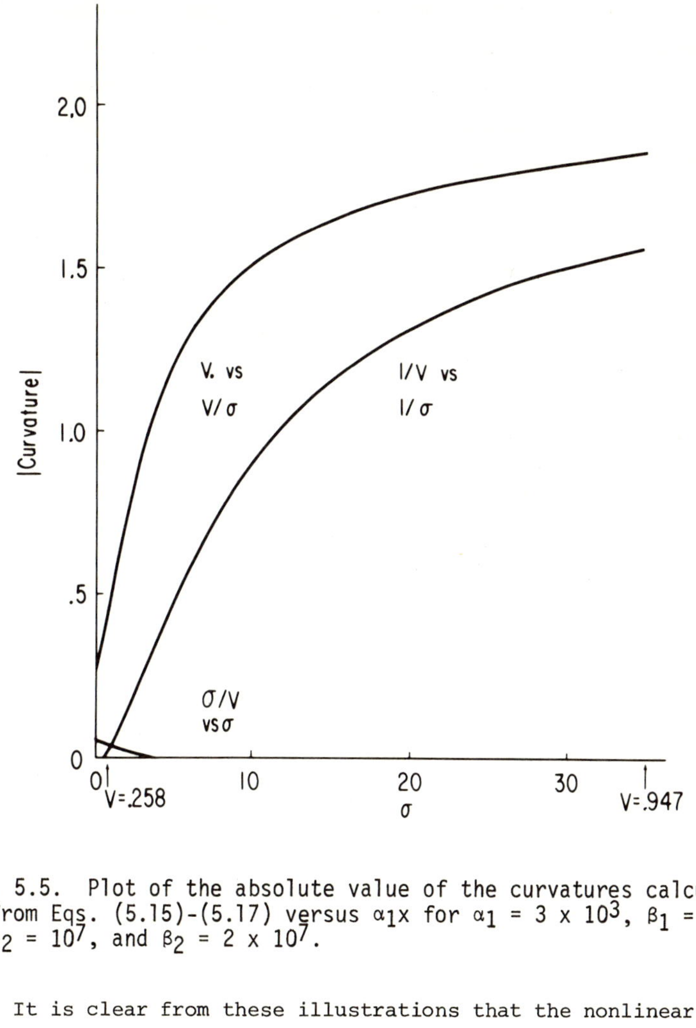

FIG. 5.5. Plot of the absolute value of the curvatures calculated from Eqs. (5.15)-(5.17) versus $\alpha_1 x$ for $\alpha_1 = 3 \times 10^3$, $\beta_1 = 10^4$, $\alpha_2 = 10^7$, and $\beta_2 = 2 \times 10^7$.

It is clear from these illustrations that the nonlinear trends originating from the degree-two model are most obvious in the v(0) versus v(0)/x type plot. Furthermore, v(0) versus v(0)/x plots can be more useful for distinguishing Eqs. (5.1) and (5.2) than statistical calculations of significance levels.[19]

Once it is decided that the degree-one model should be discarded, it becomes necessary to attempt to fit the experimental data to a different model. The simplest possibility is Eq. (5.2).

However, in order to obtain the nonlinear fit to this equation, it is necessary to have preliminary estimates for α_1, β_1, α_2, and β_2. Furthermore, these preliminary estimates must be in the neighborhood of the true values of the parameters, or there is no guarantee that a particular nonlinear fitting method will converge to the correct parameter values. These are not trivial matters, but it is beyond the scope of this text to deal with them in detail. However, if one uses the values of α_1 and β_1 obtained from the weighted fit to Eq. (5.6) together with $\alpha_2 = \alpha_1{}^2$ and $\beta_2 = \beta_1{}^2$ as the initial guesses for a nonlinear fit of the data in Fig. 5.2 (a) to Eq. (5.2), the sum of the squares of the deviations is decreased by more than an order of magnitude when the final estimates for the parameters is reached.

5.3 INTEGRATED RATE EQUATIONS AND METHODS FOR ESTIMATING INITIAL RATES

The time-honored method of obtaining initial rates is to estimate the initial tangent of P(t) or S(t) visually with a straightedge. This method includes an unnecessary step involving human judgment. As a consequence the method is vulnerable to the bias of the individual placing the straightedge and even to the potential bias introduced by the nature of the straightedge itself. We[21] have found, for example, that initial slopes estimated visually from P(t) are generally lower than the estimates obtained from graphical, numerical, or analog computer fits. Since these differences are systematically greater in experiments involving lower initial rates even when the reaction order is unchanged, visual estimates of initial rates can clearly introduce systematic errors into plots like those illustrated in Figs. 5.1 and 5.2.

The most straightforward method of fitting kinetic data to models is to simulate the model and compare the theoretical and experimental P(t) or S(t). We[21] have simulated enzyme mechanisms like the one described in Eqs. (2.1) (repeated for convenience),

$$E + S \underset{k_{-1}}{\overset{k_1}{\rightleftharpoons}} X_1 \underset{k_{-2}}{\overset{k_2}{\rightleftharpoons}} E + P \tag{5.28}$$

on an accurate electronic analog computer. $\pi(t)$ from the model and $P(t)$ from an experiment were compared in the computer. The best set of parameters was obtained by minimizing the integral of the square of the differences, $\int [\pi(t) - P(t)]^2$. In the program in Fig. 5.6 the settings of the four potentiometers ($P1 = K_1$, $P2 = 0.1K_{-1}$, $P3 = 0.1K_2$, and $P4 = K_{-2}$) that yield the minimum output from integrator A4 are the set of values of these four rate constants that gives the best fit between the model [Eqs. (5.28)] and the experimental results.

We[21] have also used an analog computer to fit experimental $P(t)$ to Eq. (3.40) (repeated for convenience) when $n = 2$:

$$t = C + \eta_0 P + \frac{\eta_1}{2} P^2 + \frac{\eta_2}{3} P^3 \tag{5.29}$$

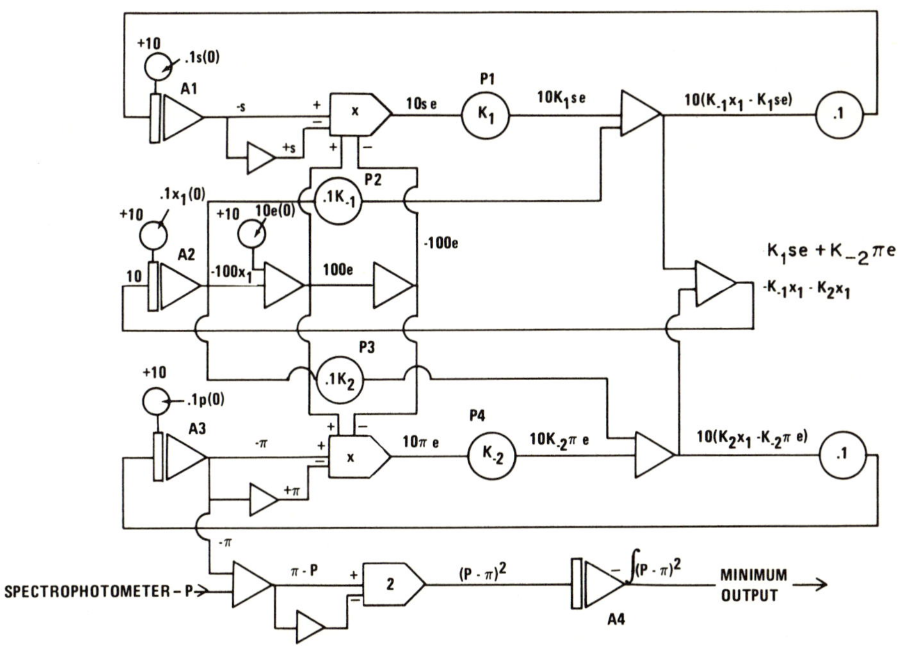

FIG. 5.6. Analog-computer program to fit experimental data to the mechanism described in Eqs. (5.28).

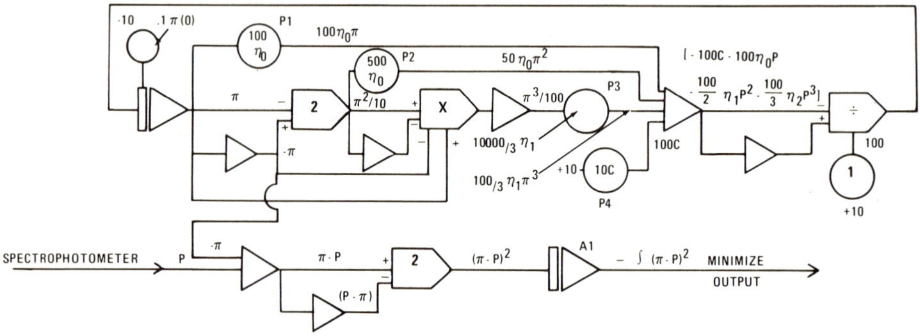

FIG. 5.7. Analog-computer program to fit experimental data to Eq. (5.29).

In the program in Fig. 5.7 the settings of the four potentiometers [$P1 = 100\eta_0$, $P2 = 500\eta_1$, $P3 = (10000/3)\eta_2$, and $P4 = 10C$] that result in the minimum output from amplifier A1 are the four constants in Eq. (5.29) that best fit the experimental data when $n = 2$.

Equation (5.29) provides a sound basis for systematic methods of fitting enzyme kinetic data to a convenient mathematical form. We[21] have developed a graphical procedure for estimating initial rates from P(t). This procedure uses Eq. (5.29), but it requires a sequence of plots of: (1) t/P^2 versus $1/P$, (2) $t - \eta_0 P$ versus P^2, (3) $t - \eta_0 P - \eta_1 P^2$ versus P^3, and (4) $P + (\eta_1/2\eta_0)P^2 + (\eta_2/3\eta_0)P^3$ versus t. This procedure is very time consuming, but it has the advantage that a computer is not required.

A much more convenient procedure is to fit P(t) directly to Eq. (3.40). We[22] have found that for most enzyme kinetic data that has been collected during the early stages of the overall reaction, Eq. (3.40) can be abridged at $n = 1$:

$$t = C + \eta_0 P + 0.5\eta_1 P^2 \tag{5.30}$$

where the higher-order terms in Eq. (3.40) have been neglected because during the early stage of the reaction, P(t) is small. The calculation of the three constants in Eq. (5.30) from P(t) data is straightforward:

$$C = \frac{\pi_1\tau_0 - \pi_2\tau_1 + \pi_3\tau_2}{\delta} \qquad (5.31)$$

$$\eta_0 = \frac{\pi_5\tau_1 - \pi_2\tau_0 - \pi_6\tau_2}{\delta} \qquad (5.32)$$

$$\eta_1 = \frac{2(\pi_3\tau_6 - \pi_6\tau_1 - \pi_4\tau_2)}{\delta} \qquad (5.33)$$

where

$$\pi_1 = \Sigma\, P^2\, \Sigma\, P^4 - (\Sigma\, P^3)^2$$

$$\pi_2 = \Sigma\, P\, \Sigma\, P^4 - \Sigma\, P^2\, \Sigma\, P^3$$

$$\pi_3 = \Sigma\, P\, \Sigma\, P^3 - (\Sigma\, P^2)^2$$

$$\pi_4 = N\, \Sigma\, P^2 - (\Sigma\, P)^2$$

$$\pi_5 = N\, \Sigma\, P^4 - (\Sigma\, P^2)^2$$

$$\pi_6 = N\, \Sigma\, P^3 - \Sigma\, P\, \Sigma\, P^2$$

$$\tau_0 = \Sigma\, t$$

$$\tau_1 = \Sigma\, tP$$

$$\tau_2 = \Sigma\, tP^2$$

$$\delta = 2\, \Sigma\, P\, \Sigma\, P^2\, \Sigma\, P^3 - N(\Sigma\, P^3)^2 + N\, \Sigma\, P^2\, \Sigma\, P^4 - (\Sigma\, P^2)^3 - (\Sigma\, P)^2\, \Sigma\, P^4$$

and N is the total number of data points collected. The variances of these parameters are

$$\text{var}(C) = \frac{\pi_1\phi}{\delta} \qquad (5.34)$$

$$\text{var}(\eta_0) = \frac{\pi_5\phi}{\delta} \qquad (5.35)$$

$$\text{var}(\eta_1) = \frac{\pi_4\phi}{\delta} \qquad (5.36)$$

where

$$\phi = \frac{\Sigma\, t^2 - C\tau_0 - \eta_0\tau_1 - \eta_1\tau_2}{N - 3}$$

If we differentiate Eq. (3.40) and solve for $\dot{P}$, we obtain

$$\dot{P} = \frac{1}{\sum_{i=0}^{n-1} \eta_i P^i} \tag{5.37}$$

During the early stage of the overall reaction when Eq. (5.30) is valid,

$$\dot{P} = \frac{1}{\eta_0 + \eta_1 P} \tag{5.38}$$

$$\ddot{P} = -\frac{\eta_1}{(\eta_0 + \eta_1 P)^3} \tag{5.39}$$

Mathematical extrapolation to the initial rate is done by setting $P = P(0) = 0$:

$$v(0) \equiv \dot{P}(0) = \frac{1}{\eta_0} \tag{5.40}$$

$$\dot{v}(0) \equiv \ddot{P}(0) = -\frac{\eta_1}{\eta_0^3} \tag{5.41}$$

If we compare Eqs. (5.40) and (5.41) with the usual expressions for the initial, quasi-steady-state rate or the initial, quasi-steady-state deceleration,

$$v(0) = \frac{V_m\, S(0)}{K_m + S(0)} \tag{5.42}$$

$$\dot{v}(0) = -\frac{V_m^2 K_m S(0)}{[K_m + S(0)]^3} \tag{5.43}$$

we find that

$$\eta_0 = \left[\frac{K_m}{S(0)} + 1\right]\frac{1}{V_m} \tag{5.44}$$

$$\eta_1 = -\frac{K_m}{V_m S(0)^2} \tag{5.45}$$

Thus, the limiting rate and the apparent Michaelis constant can be calculated directly from S(0), η_0, and η_1:

$$V_m = \frac{1}{\eta_0 - \eta_1 S(0)} \tag{5.46}$$

$$K_m = \frac{\eta_1 S(0)^2}{\eta_0 - \eta_1 S(0)} \tag{5.47}$$

Alternatively, $1/\eta_0$ could be obtained for a series of different initial conditions and then fit to a suitable function of the initial concentrations.

The details of an actual laboratory setup wherein P(t) for an enzyme-catalyzed reaction is automatically collected and fit to Eq. (5.30) appear in the flowchart in Table 5.2. The typical computer output for an actual experiment with glycerol dehydrogenase in Table 5.3 lists the calculated coefficients (A in Table 5.3 is C in Eq. (5.30), B is η_0, and C is η_1), their variances ("- OR +"), the initial rate ("RATE") and its variance ("INTERVAL"), and the initial deceleration ("ACC").

TABLE 5.2. Flowchart for collecting and analyzing enzyme kinetic data.

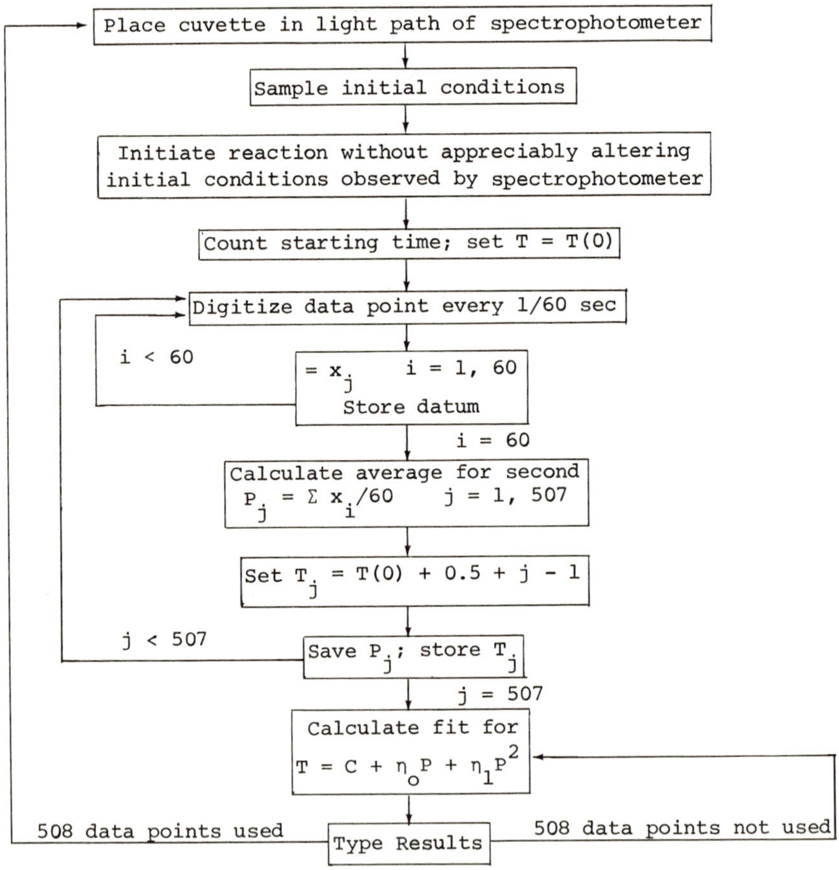

TABLE 5.3. Computer output during the collection and analysis of enzyme kinetic data.

```
DATA FROM TAPE UNIT 1 .

TYPE 3-DIGIT OCTAL BN WHERE DATA BEGINS.
061
DATA IS FROM THE CARY.

TYPE FULL SCALE OD THEN ZERO SCALE OD :
.2
0

THIS DATA WAS COLLECTED ON 06/22.
THERE ARE  +5.080000E+002 DATA POINTS NOT INC THE IC .
HOW MANY POINTS PER CALCULATION ?
120

FACTOR TO CONVERT DATA TO MICROMOLAR:
5.0

FIT FOR QUADRATIC :     T  = A + BP + CPXP

A =   -3.212625E+000   - OR   +1.269799E-001
B =   +6.171486E+002   - OR   +2.802571E+000
C =   +2.424010E+002   - OR   +1.318674E+001
RATE = +1.620355E-003   INTERVAL = +1.613030E-003   TO   +1.627746E-003
ACC =   -2.062505E-006

FIT FOR QUADRATIC :     T  = A + BP + CPXP

A =   -5.535091E+000   - OR   +4.618322E-001
B =   +6.646708E+002   - OR   +5.377828E+000
C =   +7.814675E+001   - OR   +1.339806E+001
RATE = +1.504504E-003   INTERVAL = +1.492428E-003   TO   +1.516775E-003
ACC =   -5.322567E-007

FIT FOR QUADRATIC :     T  = A + BP + CPXP

A =   -5.762249E+000   - OR   +3.628653E-001
B =   +6.761894E+002   - OR   +2.904359E+000
C =   +2.688514E+001   - OR   +4.894699E+000
RATE = +1.478875E-003   INTERVAL = +1.472550E-003   TO   +1.485254E-003
ACC =   -1.739151E-007

FIT FOR QUADRATIC :     T  = A + BP + CPXP

A =   -2.504193E+000   - OR   +3.635984E-001
B =   +6.380183E+002   - OR   +2.266759E+000
C =   +1.002623E+002   - OR   +2.973701E+000
RATE = +1.567353E-003   INTERVAL = +1.561804E-003   TO   +1.572941E-003
ACC =   -7.720908E-007

FIT FOR QUADRATIC :     T  = A + BP + CPXP

A =   -1.558523E+000   - OR   +4.632670E-001
B =   +6.284484E+002   - OR   +2.749340E+000
C =   +1.158807E+002   - OR   +3.434473E+000
RATE = +1.591220E-003   INTERVAL = +1.584289E-003   TO   +1.598212E-003
ACC =   -9.337546E-007

FINISHED WITH THAT DATA.
```

5.4 INTEGRATED FORMS OF QUASI-STEADY-STATE RATE EQUATIONS

An alternative to using the differential equation for the analysis of quasi-steady-state data is to use the integral equation relating the reactant concentrations and time. If we take advantage of the fact that the negative term in the numerator of Eq. (2.12) is negligible provided the overall reaction is sufficiently far from equilibrium,[5] we obtain the simplified quasi-steady-state differential equation for S(t) for reversible enzyme mechanisms involving one substrate and product:

$$-\dot{S}(t) = \frac{\dfrac{V_f}{K_f} S(t)}{\dfrac{S(t)}{K_f} + \dfrac{S(0) - S(t)}{K_r} + 1} \tag{5.48}$$

In Eq. (5.48) $P(t) \simeq S(0) - S(t)$, and $\dot{P}(t) \simeq -\dot{S}(t)$ if the quasi-steady-state approximation is valid, and V_f, K_f, and K_r are defined in connection with Eq. (1.14). Equation (5.48) can be put in the form Heineken and associates[23] used for the irreversible mechanism involving one enzyme intermediate:

$$\dot{x} = f(x, a) = -\mu \frac{a_1 x}{a_2 + x} \qquad [x(0) = 1] \tag{5.49}$$

In the case of the general mechanism, Eq. (5.48),

$$x = \frac{S(t)}{S(0)} \qquad \mu = \frac{E(0)}{S(0)}$$

$$a_1 = \frac{V_f}{E(0)} \frac{K_r}{K_r - K_f} \qquad a_2 = \frac{K_f}{K_r - K_f} \frac{K_r + S(0)}{S(0)}$$

Equation (5.49) is an ordinary differential equation involving two parameters. A method for the estimation of these parameters has been outlined by Heineken and associates.[23] As usual, one supposes that there is a set of observations z obtained at a number of instants t_r, $z_r = z(t_r)$, $r = 1, \ldots, R$ (during which the quasi-steady-state approximation remains valid). These observations are known functions $g(x)$ of the state variables of the system x. The state variables themselves are governed by a set of ordinary differential equations

$$\dot{x} = f(x, a, t) \qquad [x(0) = c] \tag{5.50}$$

containing a set of unknown parameters a. For the system considered here x is a single variable, $\dot{x}$ is given by the function in Eq. (5.49), and there are two parameters, a_1 and a_2. The solution of Eq. (5.50) is denoted by $x(a, c, t)$. The best choice of a is obtained by minimizing the weighted sum of squares,

$$S^2 = \sum_{r=1}^{R} \{z_r - g[x(a, c, t_r)]\}^T M_r^{-1} \{z_r - g[x(a, c, t_r)]\} \tag{5.51}$$

where M_r is the covariance matrix of the observations at times t_r.

If we have a continuous record of $S(t)$ [Eq. (5.48)], we divide it by $S(0)$, obtain the continuous observation $z(t)$, and compare it to the solution $x(t, a_1, a_2)$. If the variance of the measurement of $S(t)$ is independent of time,

$$\bar{S}^2 = \frac{1}{\sigma^2 T} \int_0^T [z(t) - x(t, a_1, a_2)]^2 \, dt \tag{5.52}$$

is the function that must be minimized.

Next we assume that $x(0) = 1$ is known with great accuracy, and that the errors in the observations of $x(t)$ are normally distributed with zero mean and covariance matrix M_r or $M(t)$. Heineken and associates[23] recommend the use of this matrix in the weighted sum of squares because it minimizes the volume of the ellipsoid of error and makes the standard deviation of linear combination of the parameters as small as possible. If these variances are sufficiently small and linearization is valid, the errors in the estimates of the parameters will also be normally distributed with zero mean and covariance matrix P. The object then is to calculate P for the parameter values that minimize the sum of the squares.

For the general case we evaluate the matrices A(t) and B(t),

$$A_{ij} = \frac{\partial f_i}{\partial x_j} \qquad B_{ik} = \frac{\partial f_i}{\partial a_k} \tag{5.53}$$

for the solution trajectory x(a*, t). The matrix D(t) is calculated by solving

$$\dot{D} = AD + B \qquad [D(0) = 0] \tag{5.54}$$

We next set $D_r = D(t_r)$ and $G_r = G(t_r)$, where

$$G_{ij} = \frac{\partial g_i[x(a^*, t)]}{\partial x_j} \tag{5.55}$$

Then if the matrix H is

$$H = \sum_{r=1}^{R} D_r^T G_r^T M_r^{-1} G_r D_r \tag{5.56}$$

the P matrix is

$$P = H^{-1} \tag{5.57}$$

For the case we are considering here [Eq. (5.49)], A and B are

$$A = \frac{\partial f}{\partial x} = -\frac{\mu a_1 a_2}{(a_2 + x)^2}$$

$$B_1 = \frac{\partial f}{\partial a_1} = -\frac{\mu x}{a_2 + x} \qquad B_2 = \frac{\partial f}{\partial a_2} = \frac{\mu a_1 x}{(a_2 + x)^2} \tag{5.58}$$

Heineken and associates [23] suggest writing Eqs. (5.54) with x as the independent variable:

$$\frac{dD_1}{dx} = \frac{a_2}{x(a_2 + x)} D_1 + \frac{1}{a_1} \qquad [D_1(1) = 0]$$

$$\frac{dD_2}{dx} = \frac{a_2}{x(a_2 + x)} D_2 + \frac{1}{a_2 + x} \qquad [D_2(1) = 0] \tag{5.59}$$

Equations (5.59) can be integrated analytically to give $D_1(x)$ and $D_2(x)$

$$D_1(x) = \frac{x\,\phi(x, a_2)}{a_1(a_2 + x)}$$
$$D_2(x) = \frac{x \ln 1/x}{a_2 + x} \qquad (5.60)$$

Since z is merely the fractional concentration, the G matrix [Eq. (5.55)] reduces to the scalar unity:

$$G = 1 \qquad (5.61)$$

The matrix h therefore has the elements

$$h_{11} = \frac{\alpha(\xi, a_2)}{\sigma^2 a_1^2 \phi} \qquad h_{12} = h_{21} = -\frac{\beta(\xi, a_2)}{\sigma^2 a_1 \phi} \qquad h_{22} = \frac{\gamma(\xi, a_2)}{\sigma^2 \phi} \qquad (5.62)$$

where

$$\alpha(\xi, a_2) = \int_\xi^1 \frac{x[\phi(x, a_2)]^2\,dx}{a_2 + x} \qquad (5.63)$$

$$\beta(\xi, a_2) = \int_\xi^1 \frac{x\,\phi(x, a_2)\ln(1/x)\,dx}{a_2 + x} \qquad (5.64)$$

$$\gamma(\xi, a_2) = \int_\xi^1 \frac{x[\ln(1/x)]^2\,dx}{a_2 + x} \qquad (5.65)$$

If $\delta^2 = \alpha\gamma - \beta^2$, then

$$h^{-1} = \sigma^2 \phi \begin{vmatrix} \dfrac{a_1^2 \gamma}{\delta^2} & \dfrac{a_1 \beta}{\delta^2} \\[2ex] \dfrac{a_1 \beta}{\delta^2} & \dfrac{\alpha}{\delta^2} \end{vmatrix} \qquad (5.66)$$

Heineken and associates[23] point out that if the observations are equally spaced on the interval (0, T), and if the number of observations is large enough,

$$P \simeq \frac{h^{-1}}{R} \tag{5.67}$$

Equations (5.66) and (5.67) therefore provide a basis for the calculation of the variances in the estimates of a_1 and a_2. The variance of a_1 comes from the first row and first column of Eq. (5.66),

$$\text{var}(a_1) \equiv \sigma_1^{\ 2} = \frac{1}{R} \frac{\sigma^2 \phi a_1^{\ 2} \gamma}{\delta^2} \tag{5.68}$$

The variance of a_2 comes from the second row and second column:

$$\text{var}(a_2) \equiv \sigma_2^{\ 2} = \frac{1}{R} \frac{\sigma^2 \phi \alpha}{\delta^2} \tag{5.69}$$

The correlation coefficient between the estimation errors is

$$C = \frac{\beta}{\sqrt{\alpha\gamma}} \tag{5.70}$$

Heineken and associates[23] define the quantities

$$\zeta_1 = \frac{\sigma_1 \sqrt{R}}{a_1 \sigma} = \frac{\sqrt{\phi\gamma}}{\delta} \tag{5.71}$$

$$\zeta_2 = \frac{\sigma_2 \sqrt{R}}{a_2 \tau} = \frac{\sqrt{\phi\alpha}}{a_2 \delta} \tag{5.72}$$

These quantities represent the ratio of the standard error in the parameters to the standard error in the observations.

In Fig. 5.8 ζ_1 [Fig. 5.8 (a)] or ζ_2 [Fig. 5.8 (b)] is plotted versus ξ for various values of a_2. These figures illustrate that there is a minimum ζ for a particular value of x. This means that for a fixed number of equally spaced observations, the greatest

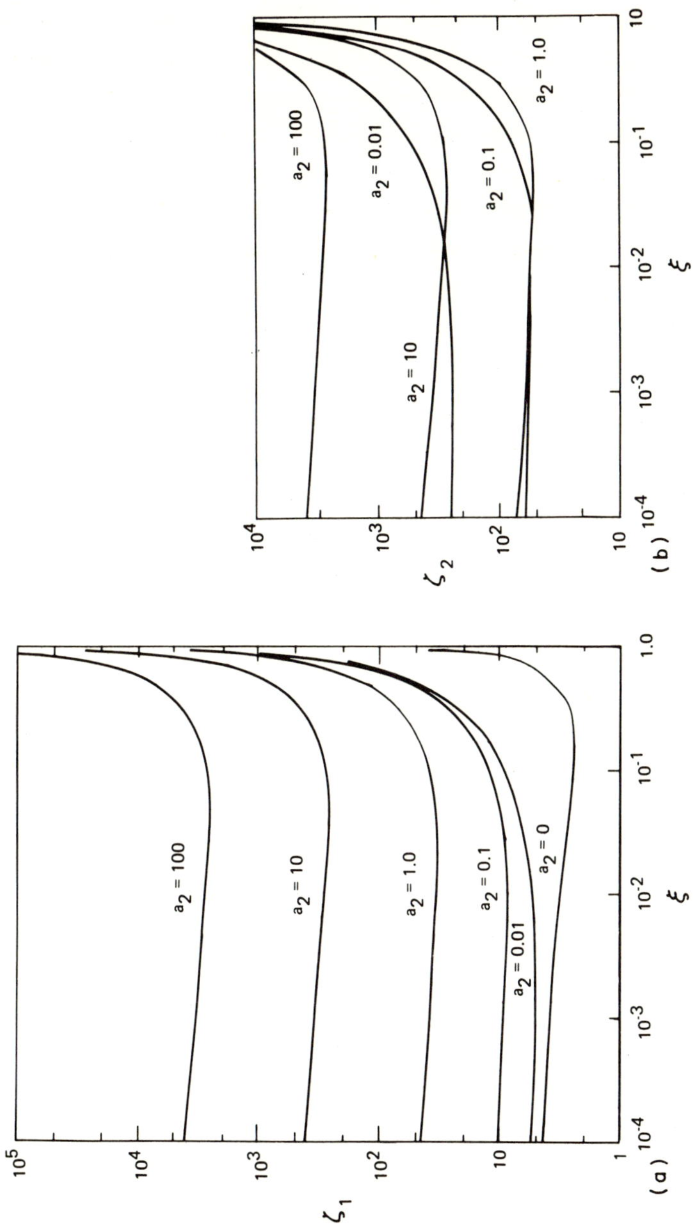

FIG. 5.8. Plots of (a) ζ_1 and (b) ζ_2 versus ζ.[20]

accuracy for the estimation of a_1 will be obtained at some particular extent of the overall reaction. Figure 5.8 (a) indicates that a_1 will be more accurately determined for smaller values of a_2, but Fig. 5.8 (b) illustrates that decreasing a_2 does not continually improve the accuracy of estimates of itself. Thus, increasing the initial substrate concentration always improves the accuracy of a_1, but not a_2. This is related to the common experience of enzymologist that it is much easier to determine a_1 accurately than a_2. A value of $a_2 = 0.1$ gives almost as good an estimate of a_2 and a much better estimate of a_1; this illustrates the advantage of using a relatively high initial substrate concentration.

Figure 5.8 also illustrates difficulties that can be encountered if one uses unsuitable experimental conditions. For example, if the a-priori estimate of the Michaelis constant is off by a factor of a hundred, experiments involving even a hundred observations could still give standard errors for a_1 and a_2 in the range of 250σ.

Figure 5.9[23] illustrates how one can use the results in Fig. 5.8 to assist in the design of a kinetic experiment with an enzyme. Suppose, for example, that one desires a_1 and a_2 with an accuracy of $\sigma_1/a_1 = 2\sigma$ and $\sigma_2/a_2 = 10\sigma$ (point A in Fig. 5.9). The effect of taking R points is to move the whole diagram downward and to the left by a distance equal to the square root of R. The first point on the diagram to encounter A is point B, where the ratio of the axes is about 5. Thus, in the most favorable circumstances, 5^2 or twenty-five experimental points are required. Furthermore, since B is between the lines for $a_2 = 1.0$ and $a_2 = 0.1$, but closest to the line for $a_2 = 0.1$, the best initial substrate concentration would be in the range of ten times the a-priori estimate for the Michaelis constant.

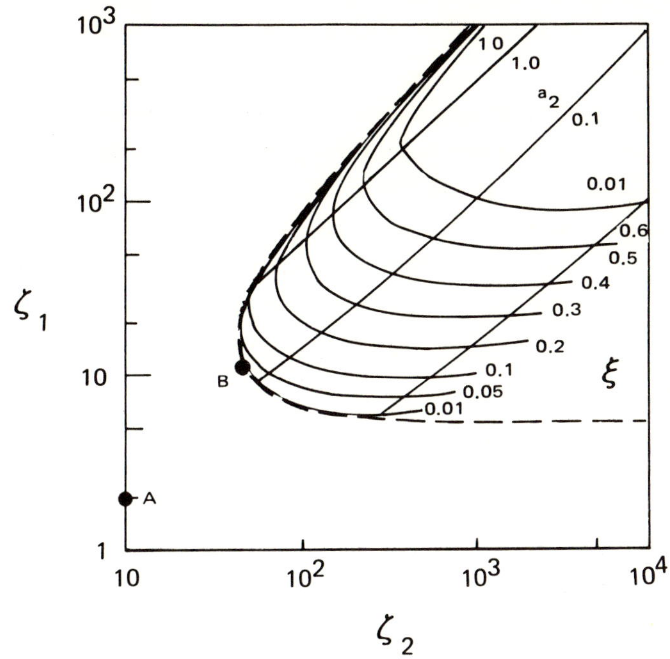

FIG. 5.9. Plot of ζ_1 versus ζ_2 for constant values of ξ and a_2.[20]

5.5 INTEGRATED FORMS OF COMPLETE RATE EQUATIONS

The University College London Enzyme Simulator (see Chapter IV) can be used to optimize the values of the rate constants in the stipulated mechanism. If one or more of the rate constants in a reaction record provided to the kinetic translator are *stipulated* (not implied) to be zero, the value of each such rate constant will be optimized to fit a set of experimental data points. The data are expected immediately after the end of the KTL list designator. The first card after the KTL list describes the format of the data. The data entries consist of up to 118 records. Each of the data records contains the time at which the datum was collected followed by the value of each of the variables mentioned in the format card. The maximum number of variables for which data can be provided is

12. The data list is terminated by a "star-star-blank-comma" (** ,) record.

After the data list has been read, control is passed to the SCL interpreter. The commands listed in Table 4.2 can be used to provide initial guesses for the rate constants, to scale the variables, and to otherwise control the execution of the optimization and simulation procedures, and to stipulate the nature of the output.

Table 5.4 is a list of the contents of 20 of the 135 punched cards that were used as input to the University College London Enzyme Simulator for an optimization run. The first card is the comment, "OPTIMIZATION". The second card is the first chemical reaction in the mechanism. 3.11×10^{-4} concentration units of a substance called "El" and 10.3 units of a substance called "Sl" are initially present. El and Sl combine reversibly to form a substance called "Cl". Cl is initially absent. The rate constants for the forward (K_1) and reverse (K_2) reactions are stipulated to be zero, so both will be optimized. The third card is the last chemical reaction in the mechanism. Cl is converted to a substance called "Pl" in a step that also regenerates El. Pl is initially absent. The rate constants for the forward (K_3) and reverse (K_4) reactions are stipulated to be zero, so both will be optimized. The fourth card indicates the end of the KTL list. The fifth card is the data format card. Time, the first number in the data records, will be followed by one variable called "Pl" in the reaction records. The next 118 cards each list a time and the corresponding value for Pl. The 124th card indicates the end of the data. The 125th card contains the initial guess for K_1, the 126th card the initial guess for K_2, the 127th card the initial guess for K_3, and the 128th card the initial guess for K_4. The 129th and 130th cards set the maximum values for the four variables, and the 131st card scales the data points for Pl to the appropriate maximum. The 132nd card produces an optimization run, and cards 133 and 134 cause graphs of the experimental variables and the variables calculated from the optimized rate constants.

TABLE 5.4. Input to optimize the individual rate constants in Eqs. (2.1) for the University College London Enzyme Simulator.

```
1 2 3 4 5 6 7 8 9 10 11 12 13 14 15 16 17 18 19 20 21 22 23 24 25 26 27 28 29 30 31 32 33 34 35 36 37 38 39 40 41 42 43 44 45
* ØPTIMIZATIØN ,
3.11E-4 E1 + 10.3 S1 = C1     /0/0 ,
C1 = E1 + P1                  /0/0 ,
** ,
TIME      P1 ,
0         .001 ,
2.271     .011 ,
```

115 MORE CARDS

```
110.084   .194 ,
** ,
K1 2.3 ,
K2 100 ,
K3 100 ,
K4 2.0 ,
MAX 3.11E-4 E1 C1 ,
MAX 10.3 S1 P1 ,
SCALE ,
BEGIN ,
GRAPH P1 .2 P ,
BEGIN ,
STØP ,
```

In Table 5.5 appears the output generated from the input in Table 5.4. The first line is the comment. The next two lines are the reaction numbers, the rate-constant subscripts, and the initial conditions for each of the two reversible chemical reactions. The fifth through eighth lines indicate the total number of chemical reactions, the total number of forward and reverse steps, the total number of variables, and whether or not syntax errors were found. The ninth line, "SQR", indicates the sum of the squares of the deviations for the first optimization of the rate constants. Lines 10 through 13 indicate the initial guesses for the four rate constants and their standard errors. Lines 18 - 21 indicate the final optimized constants and their standard errors. The graph in the next 118 lines are the experimental P1 ("1") and the calculated P1

("P") plotted versus time. The last line, "SQR", indicates the sum of the squares of the deviations for the final optimized values of

TABLE 5.5. Output for the optimization of the rate constants in Eqs. (2.1) from the University College London Enzyme Simulator.

```
                    ØPTIMIZATIØN ,
     1        1      2 3.11E-4 E1 + 10.3 S1 = C1                      / 0 / 0 ,
     2        3      4 C1 = E1 + P1                                   / 0 / 0 ,
                  *****END EQUATIØN LIST
  EQUATIØNS    =       2
  REACTIØNS    =       4
  VARIABLES    =       4
  ERRØRS =         0

 SQR =   .2720E 00
  K    1       .2300E 01      1.00
  K    2       .1000E 02      1.00
  K    3       .1000E 02      1.00
  K    4       .2000E 01      1.00
  K    1           2.7846 ,        .3997E-01        1
  K    2          10.2995 ,        .5401E-01        2
  K    3          10.1898 ,        .2140E 00        3
  K    4           6.1819 ,        .2915E-01        4
  K    1           2.8974 ,        .2445E 00        1
  K    2          10.8737 ,        .2813E-01        2
  K    3          10.2290 ,        .3887E-01        3
  K    4           8.3165 ,        .2825E-01        4
       .0000 .
       .0000 .
      2.2710 .P1
      3.2080 .P1
      4.1460 .P 1
      5.0830 . P1
      6.0210 . P 1
      6.9580 .  P1
      7.8960 .  P1
      8.8330 .   P1
      9.7710 .   P1
     10.7080 .   P 1
     11.6460 .    P1
     12.5830 .    P 1
     13.5210 .     P1
     14.4580 .     P1
     15.3960 .      P1
     16.3330 .      P1
     17.2710 .      P 1
     18.2080 .       P1
     19.1460 .       P1
     20.0830 .        P1
     21.0210 .        P1
     21.9580 .         P1
     22.8960 .         P1
     23.8330 .         P 1
     24.7710 .          P1
     25.7080 .          P1
     26.6460 .           P1
     27.5830 .           P1
     28.5210 .            P1
     29.4580 .            P1
```

TABLE 5.5 (continued)

```
30•3960 •          P 1
31•3340 •           P1
32•2710 •           P 1
33•2090 •            P1
34•1460 •            P1
35•0840 •             P1
36•0210 •             P1
36•9590 •             P1
37•8960 •              1
38•8340 •             P1
39•7710 •               1
40•7090 •              P1
41•6460 •               P1
42•5840 •               P1
43•5210 •               P1
44•4590 •                P1
45•3960 •                P1
46•3340 •                 P1
47•2710 •                 P1
48•2090 •                 P1
49•1460 •                  P1
50•0840 •                  P1
51•0210 •                   P1
51•9590 •                   P1
52•8960 •                    P1
53•8340 •                    P1
54•7710 •                    P1
55•7090 •                      1
56•6460 •                     P1
57•5840 •                       1
58•5210 •                      P1
59•4590 •                      P1
60•3960 •                        1
61•3340 •                       P1
62•2710 •                         1
63•2090 •                         1
64•1460 •                          1
65•0840 •                          1
66•0210 •                         P1
66•9590 •                           1
67•8960 •                          P1
68•8340 •                            1
69•7710 •                            1
70•7090 •                           P1
71•6460 •                             1
72•5840 •                             1
73•5210 •                              1
74•4590 •                              1
75•3960 •                             P1
76•3340 •                               1
77•2710 •                               1
78•2090 •                                1
79•1460 •                                1
80•0840 •                               P1
81•0210 •                                 1
81•9590 •                                 1
82•8960 •                                  1
83•8340 •                                  1
84•7710 •                                  1
85•7090 •                                   1
```

TABLE 5.5 (continued)

86•6460	•	1
87•5840	•	1
88•5210	•	1
89•4590	•	P1
90•3960	•	1
91•3340	•	1
92•2710	•	1P
93•2090	•	1
94•1460	•	1P
95•0840	•	1P
96•0210	•	1
96•9590	•	1P
97•8960	•	1
98•8340	•	1P
99•7710	•	1P
100•7090	•	1
101•6460	•	1
102•5840	•	1
103•5210	•	1
104•4590	•	1
105•3960	•	1
106•3340	•	1
107•2710	•	1
108•2090	•	1
109•1460	•	1
110•0840	•	1

SQR = •2582E-01

the rate constants. The execution time for the input in Table 5.4 was 60.2 sec on a Sigma V computer.

The University College London Enzyme Simulator and Optimization Program will fit *any* set of data to *any* mechanism provided in a KTL list. Even data that do not contain information about a specific aspect of a mechanism will lead to a value for any constant in the mechanism. For example, quasi-steady-state kinetic data for enzymes contain information about the quasi-steady-state phase of the overall reaction. Quasi-steady-state parameters like Michaelis constants and maximum velocities can be estimated from such data, but the individual rate constants even in the simple mechanism described by Eqs. (2.1) cannot all be obtained from forward reaction data alone. An optimization program does not "know" that specific information about a given mechanism is or is not available from a particular set of experimental data. The optimization is therefore carried out as though the constants to be optimized were well defined by the input. If they are not, the results may reflect the initial guesses rather than the actual values for ill-defined parameters because a range of values, and perhaps even several different ranges of values, of a poorly-defined constant may satisfy the fitting criteria. Under these conditions the optimization program uses the initial guess to locate the nearest local value of the constant that "optimizes" the fit.

In Table 5.6 appear values for "optimized" rate constants and their standard error for six ranges of the same experimental data and two different initial guesses. The optimizations were obtained from the KTL list and SCL commands in Table 5.4. The experimental data [P(t)] were collected beginning 1 sec after mixing the enzyme and its substrates but before the overall reaction was near equilibrium. Since E(0)/S(0) is 3×10^{-5}, we expect quasi-steady-state data. K_3 is the only individual rate constant usually available from such data. The "optimized" values of the other constants were always found near their respective initial guesses. The "correct" values of these ill-defined rate constants were not the values obtained from the optimization program.

The problem of local optimization of ill-defined constants is not due to a fault in the optimization procedure. The program does what it is told to do by the user. The user must be aware of the type of information available from the experimental data. If the desired information is not available from, for example, quasi-steady-state data, it is up to the experimentalist to design a better experiment. Otherwise he must be satisfied with kinetic parameters like Michaelis constants and maximum velocities rather than individual rate constants.

TABLE 5.6. Optimized values for the rate constants in Eq. (2.1) found for different initial guesses and different ranges of the same experimental data.

Initial guess	Time range (seconds) for data	K_1	S.E.[a]	K_2	S.E.	K_3	S.E.	K_4	S.E.	Sum of squares	K_f	K_r
A:	1-111	2.900	0.245	10.874	0.0281	10.229	0.0389	8.316	0.0282	0.0258	7.28	2.54
K_1 = 2.3	112-221	2.522	0.0585	10.682	0.00795	10.233	0.0220	3.759	0.0142	0.00598	8.29	5.56
K_2 = 10.0	222-331	2.582	0.101	10.626	0.00376	10.230	0.0143	3.147	0.0135	0.00349	8.08	6.63
K_3 = 10.0	332-441	2.467	0.0390	10.662	0.00551	10.242	0.0204	2.738	0.00932	0.00106	8.47	7.63
K_4 = 2.0	1-220	2.577	0.0752	10.679	0.0148	10.236	0.0238	4.670	0.0190	0.0135	8.12	4.48
K_f = 8.7	1-437	2.525	0.0507	10.659	0.00522	10.241	0.0131	3.118	0.00911	0.00409	8.28	6.70
K_r = 10.0												
B:	1-111	5.960	0.306	32.205	0.0268	10.246	0.0394	9.951	0.0293	0.0217	7.12	4.27
K_1 = 4.6	112-221	5.427	0.195	31.981	0.00694	10.253	0.0305	6.363	0.0313	0.00612	7.78	6.64
K_2 = 30.0	222-331	5.546	0.211	32.376	0.0119	10.241	0.0245	3.976	0.00263	0.00256	7.68	10.7
K_3 = 10.0	332-441	5.563	0.286	31.958	0.000048	10.254	0.0310	4.823	0.0122	0.00038	7.59	8.75
K_4 = 4.0	1-220	4.450	0.113	32.005	0.0112	10.249	0.0183	6.266	0.0162	0.00956	7.75	6.74
K_f = 8.7	1-437	5.616	0.139	32.374	0.00467	10.248	0.0176	4.827	0.00916	0.00376	7.59	8.83
K_r = 10.0												

[a]Standard error.

CHAPTER 6
THEORY OF CONTROL

6.1 INTRODUCTION

The kinetic or dynamic behavior of a system of n interconnected chemical reactions can be represented by

$$\frac{dx_i}{dt} = X_i(t, x_1, \ldots, x_n; \alpha_1, \ldots, \alpha_m) \quad [i = 1, \ldots, n] \qquad (6.1)$$

where t is the time, the x's are the variable chemical concentrations, the α's are constant parameters, and the X's are generally nonlinear functions. If the right-hand side of the differential equations do not contain time, the system is said to be *autonomous*, and we may write

$$\dot{x}_i = f_i(x_1, \ldots, x_n) \qquad [i = 1, \ldots, n] \qquad (6.2)$$

where the f's are nonlinear functions containing the parameters α.

The *stationary solutions* of Eqs. (6.2) are the roots of the algebraic expressions

$$f_i(x_1{}^*, \ldots, x_n{}^*) = 0 \qquad [i = 1, \ldots, n] \qquad (6.3)$$

Each set of $x_i{}^*$ satisfying Eqs. (6.3) is referred to as a *singularity* or a *steady-state solution* of Eqs. (6.2). In chemical systems we are interested in sets of real, nonnegative $x_i{}^*$.

If we introduce the change of variables

$$u_i = x_i - x_i^* \qquad [i = 1, \ldots, n] \tag{6.4}$$

we obtain

$$\dot{u}_i = U_i(u_1, \ldots, u_n) \qquad [i = 1, \ldots, n] \tag{6.5}$$

such that $U_i = 0$ when $u_i = 0$ for all $i = 1, \ldots, n$. The singularity $u_i = 0$ is called the undisturbed motion of Eqs. (6.5). This undisturbed motion is called *stable* if, for all disturbances, Δu_i satisfying the condition

$$\sum_{i=1}^{n} \Delta_i^2 \leq H \tag{6.6}$$

the motion of the differential equations is such that

$$\sum_{i=1}^{n} u_i^2 < A \tag{6.7}$$

for all possible values of the variables in the real (positive) time domain. In other words, if disturbances limited to an H vicinity, do not result in variable excursions beyond an A vicinity, the motion of the system is stable.

The nature of the stability of the motion in the A vicinity can be twofold. The motion will be stable in the sense either that inequalities (6.6) and (6.7) are satisfied, or that in addition,

$$\lim_{t \to \infty} u_i(t) = 0 \tag{6.8}$$

If Eq. (6.8) is satisfied, the stability of the motion is called *asymptotic*.

6.2 THE LYAPUNOV DIRECT METHOD

One of the most important points of the Lyapunov method is that it provides a basis for the calculation of H in Eq. (6.6). The magnitude of this number depends upon the form of the so-called *Lynapunov V function* and the parameters of the differential equations that describe the system. After H has been calculated, it is possible to establish the size of the region in which the stability is assured. If H is not sufficiently large to include all of the phase-space accessible to the variables, we speak of *stability in the small* or *local stability*. If H is global, we speak of *stability in the large* or *global stability*.

The Lyapunov direct method involves the construction of a function of the variables whose Eulerian derivative has certain properties that ensure stability or asymptotic stability. If the Lyapunov function is unbounded and the properties of the derivative ensure asymptotic stability, the stability is asymptotic and global.

A function is called *sign invariant* if it is always of the same sign or zero. An example of a positive-invariant function is

$$V = (u_1 + u_2)^2 + u_3^2 \tag{6.9}$$

This function is positive for all values of the variables except $u_1 = -u_2$, $u_3 = 0$, or $u_1 = u_2 = u_3 = 0$, in which cases it is zero. A sign-invariant function that is zero only at the origin is called *sign definite*. An example of a positive-definite function is

$$V = u_1^2 + u_2^2 + u_3^2 \tag{6.10}$$

If V is a sign-definite function, then V = C = constant represents a one-parameter family of curves in the phase space of the variables. If C is sufficiently small, each of these curves will be closed, and they must contain the origin. The two-dimensional representation of this situation is illustrated in Fig. 6.1. Closure of these surfaces is ensured[24] if the Lyapunov V function approaches infinity as

$$\sum_{i=1}^{n} u_i^2 \to \infty \tag{6.11}$$

It is clear from Fig. 6.1 that the magnitude of V defines the domain of the phase space under investigation. If V is unbounded, the phase space enclosed by the function will be global.

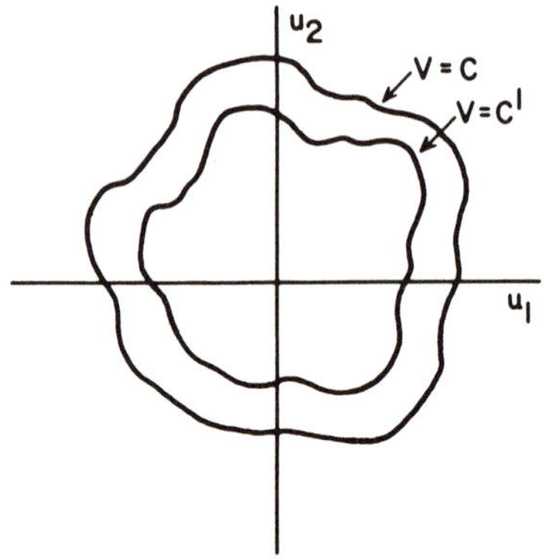

FIG. 6.1. Two-dimensional representation of a Lyapunov function.

Regardless of the size of the region defined by a Lyapunov function, the surfaces surrounding the origin intersect all paths that lead from the origin to infinity. Thus, if all motion in the phase space across the whole family of an unbounded Lyapunov function is from outside the surfaces into the region enclosed by the surfaces, the motion must possess global asymptotic stability. If, on the other hand, this motion tends toward a finite closed surface, the motion possesses stability, but the stability is not asymptotic.

Theorem I (Lyapunov): *A motion is stable if it is possible to find a sign-definite function V having a Eulerian derivative that is sign invariant and of sign opposite to that of V.*

Theorem II (Lyapunov): *A motion is asymptotically stable if it is possible to find a sign-definite function V having a Eulerian derivative that is sign definite and of sign opposite to that of V.*

To prove Theorem II, assume that V is positive-definite. The total derivative with respect to time is

$$\frac{dV}{dt} = \text{grad } V \cdot U = \sum_{i=1}^{n} \frac{\partial V}{\partial u_i} \dot{u}_i = \sum_{i=1}^{n} \frac{\partial V}{\partial u_i} U_i \tag{6.12}$$

Since the $\partial V/\partial u_i$ are proportional to the directional cosines of the normal N to the surface V = C, Eq. (6.12) can be written as[25]

$$\dot{V} = \left[\sum_{i=1}^{n} \left(\frac{\partial V}{\partial u_i}\right)^2\right]^{\frac{1}{2}} \sum_{i=1}^{n} U_i \cos Nu_i \tag{6.13}$$

where $\Sigma U_i \cos Nu_i$ represents the projection of the velocity of the representative point R on the normal to the V = C surface. Since $\Sigma U_i \cos Nu_i < 0$ means that R moves along trajectories that intersect the family of surfaces inward, R cannot remain outside or on any surface V = C ($C \neq 0$). Consequently, as time approaches infinity, the motion will approach the origin and will thus be asymptoically stable. A two-dimensional representation of this situation appears in Fig. 6.2.

The proof ot Theorem I is identical. However, the conditions of Theorem I permit $\Sigma U_i \cos Nu_i = 0$; this means that R can remain on some surface V = C. In this case the motion is stable, but the stability is not asymptotic.

These two Lyapunov theorems are powerful tools in studies of stability properties of nonlinear control systems. Often it is not a trivial problem to construct an appropriate Lyapunov function

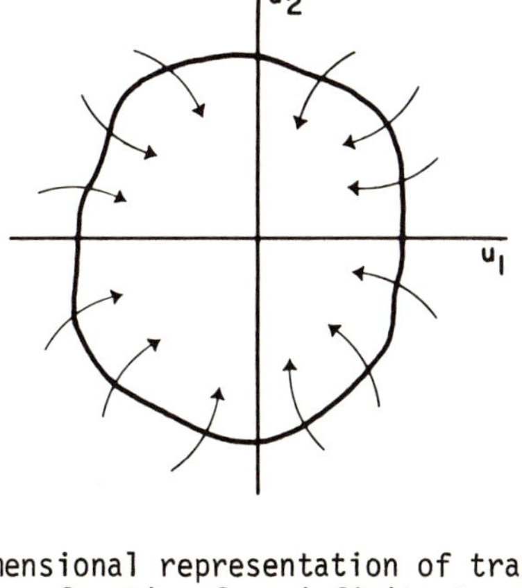

FIG. 6.2. Two-dimensional representation of trajectories intersecting a Lyapunov function from infinity toward the origin.

for a particular set of nonlinear differential equations. However, if a global Lyapunov function can be constructed, important conclusions not otherwise apparent can often be deduced about the dynamics of the control system under investigation.

The Lyapunov theorems can also be used to deduce the well-known stability properties of linearized systems. If we develop the right-hand sides of Eqs. (6.5) into power series that converge in the neighborhood of the origin, we obtain

$$\dot{u}_i = \sum_{i=1}^{n} a_{ij}u_j + U_{2i}(u_1, \ldots, u_n) \qquad [i = 1, \ldots, n] \tag{6.14}$$

where $a_{ij} = \partial U_i/\partial u_j$ evaluated at $u_j = 0$, and the nonlinear functions U_{2i} do not contain terms of order less than two. There is therefore a neighborhood of the origin wherein neglecting the U_{2i} introduces an error that is sufficiently small that it is acceptable. If we restrict the analysis (and the magnitude of the variables!) to this neighborhood, the *abridged equations*

$$\dot{u}_i = \sum_{i=1}^{n} a_{ij}u_j \qquad [i = 1, \ldots, n] \tag{6.15}$$

are an acceptable approximation for the system under investigation. A linear one-to-one transformation provides the *canonical equations*

$$\dot{z}_i = \mu_i z_i \qquad [i = 1, \ldots, n] \tag{6.16}$$

where the μ are the roots of the characteristic equation for Eq. (6.15):

$$|a_{ij} - \mu\delta_{ij}| = 0 \tag{6.17}$$

where $\delta_{ij} = 0$ for all $i \neq j$, but $\delta_{ii} = 1$. The solutions of Eq. (6.14), $\mu_1, \ldots, \mu_n$, are commonly called the *characteristic exponents* of the linearized differential equations.

Now consider the positive-definite function

$$V = \frac{1}{2}\sum_{i=1}^{n} z_i^2 \tag{6.18}$$

The total derivative with respect to time is

$$\dot{V} = \sum_{i=1}^{n} \frac{\partial V}{\partial z_i} \dot{z}_i = \sum_{i=1}^{n} \mu_i z_i^2 \tag{6.19}$$

which is negative-definite iff all the μ's are negative. (The abbreviation "iff" means "if and only if." This usage was originated by P. R. Halmos.[26]) Extending these results to the case when the μ's are conjugate-complex, we obtain the additional Lyapunov theorems:

> Theorem III (Lyapunov): *If the real parts of all the characteristic exponents are negative, the undisturbed motion is asymptotically stable in the neighborhood of the origin regardless of the terms in U_{2i}.*
>
> Theorem IV (Lyapunov): *If among the roots of the characteristic equation there is at least one with a positive real part, the undisturbed motion is unstable.*
>
> Theorem V (Lyapunov): *If one or more of the characteristics exponents is purely imaginary, the terms in U_{2i} can be chosen so that the system will be either stable or unstable.*

These results have been used for many years in studies of linearized control systems. The simplicity with which they can be deduced from the Lyapunov general theory illustrates the power and elegance of this method.

6.3 LIMIT CYCLES

In 1881 Poincaré[27] showed that differential equations like Eqs. (6.5) can sometimes result in a special type of solution represented by closed surfaces in the phase space. Poincaré called these closed trajectories *limit cycles*. These closed curves are distinguished from the closed trajectories originating from linear differential equations or conservative systems by the fact that limit cycles are *isolated* closed trajectories. Thus, the radius of a limit cycle is the same for a whole range of initial conditions.

In fact, every trajectory in the neighborhood of a stable limit cycle winds toward the limit cycle as time approaches infinity. In a trace of a variable versus time, limit-cycle behavior appears as a sustained nonlinear oscillation.

Limit cycles can be either stable or unstable. Thus, stable and unstable limit cycles represent asymptotically stable or totally unstable states of sustained oscillations in exactly the same manner as stable and unstable singularities represent stable or unstable states of constant flow. The two-dimensional behavior of trajectories in the neighborhood of stable and unstable limit cycles is illustrated in Fig. 6.3.

A limit cycle may surround a singularity that is either stable or unstable. If the singularity is unstable and the surrounding limit cycle stable, sustained oscillations will arise spontaneously (for example, due to fluctuations) from the unstable state of rest. This situation, commonly referred to as *soft excitation*, is illustrated for the two-dimensional case in Fig. 6.4 (a). On the other hand, if the singularity is stable and the surrounding limit cycle unstable, small perturbations from the state of rest will not result in sustained oscillations. However, if a stable limit cycle

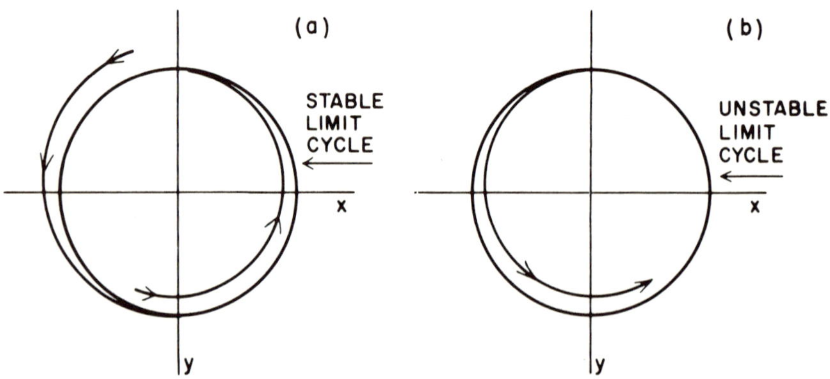

FIG. 6.3. Two-dimensional representation of (a) stable and (b) unstable limit cycles.

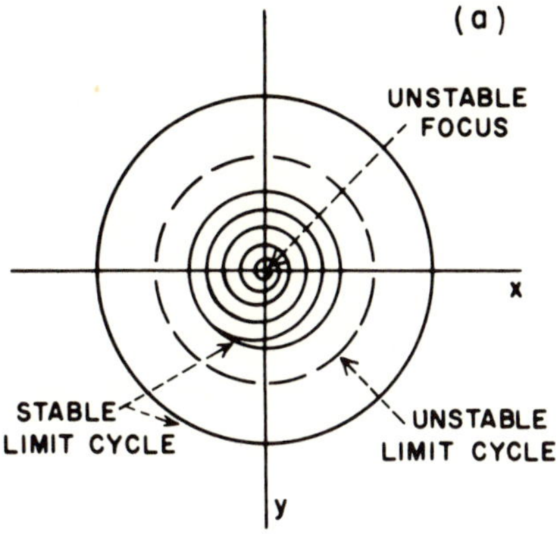

FIG. 6.4. Two-dimensional representation of alternating stable and unstable limit cycles surrounding (a) stable and (b) unstable singularities.

exists beyond the unstable cycle, sustained oscillations can arise if the state of rest is subject to a sufficiently large perturbation; this perturbation must be large enough to move the variables to the phase space outside the unstable limit cycle. This situation is commonly called *hard excitation*; the two-dimensional case is illustrated in Fig. 6.4 (b).

Figure 6.4 (b) illustrates how trajectories originating in one area of the phase space can approach a particular state of sustained oscillations, while trajectories originating elsewhere may approach a different state of oscillations, and trajectories originating yet elsewhere approach a nonoscillatory, stationary state. If the initial conditions are in the phase space between (1) a stable singularity and an unstable limit cycle, (2) an unstable and a stable limit cycle, or (3) a stable and an unstable limit cycle, the trajectories will wind (1) toward a stable stationary state or (2) and (3) toward the nearest stable limit cycle. If multiple stable limit cycles exist, different ranges of initial conditions will lead to different oscillatory regimes (for example, sustained oscillations with different amplitudes, or relaxation versus nonrelaxation oscillations). Thus, widely different initial conditions can lead to identical oscillatory regimes when the variables originate in a well-defined neighborhood of a stable limit cycle. On the other hand, initial conditions that are only slightly different but on opposite sides of an unstable limit cycle will result in completely different oscillatory regimes.

According to our definition of stability [inequalities (6.6) and (6.7)], systems exhibiting sustained oscillations are stable in the sense that they remain within a well-defined domain of the phase space. However, the stability of oscillating systems is not asymptotic toward the origin unless the oscillations are damped. Thus, if it can be shown that a system possesses global asymptotic stability (for example, with Theorem II), it follows that the system variables cannot experience sustained oscillations. However, if it can only be shown that the system possesses global stability (for example, with Theorem I), the variables could nevertheless exhibit limit-cycle behavior. Further, if it can only be shown that the system possesses local asymptotic stability (for example, with Theorem III), the system could possess surrounding unstable and stable limit cycles. Finally, even if one can show that a bounded system is unstable (for example, with Theorem IV), the system need not possess a limit cycle.[28]

Since limit cycles are the result of nonlinear terms in differential equations, it follows that one cannot decide from the nature of singular points that a surrounding limit cycle surely must exist. However, even if sufficient conditions cannot be deduced, it is possible to arrive at some necessary conditions for the existence of limit cycles. The most important of these, applicable only to systems involving two variables, is the negative criterion of Bendixson:[29]

> Theorem VI (Bendixson): *If the value of* $(\partial U_1/\partial u_1) + (\partial U_2/\partial u_2)$ *does not change sign within a region of the phase space, no closed trajectory can exist within this region.*

CHAPTER 7

CHEMICAL KINETICS OF MULTICOMPONENT SYSTEMS WITHOUT FEEDBACK

7.1 KINETICS OF LINEARIZED, CLOSED CHEMICAL SYSTEMS

In a remarkable paper published twenty years ago,[30] John Hearon pointed out certain restrictions that thermodynamic principles impose on chemical systems described by linearized rate equations. If we consider a small enough interval for the variables in Eq. (6.2), we may linearize the f_i

$$f_i(x_1, \ldots, x_n) = \sum_{j=1}^{n} a_{ij}x_j + f_{2i}(x_1, \ldots, x_n) \qquad [i = 1, \ldots, n] \qquad (7.1)$$

by neglecting the terms of order two or greater [i.e., the terms in f_{2i} in Eqs. (7.1)]. The linearized rate equations are

$$\dot{x}_i = \sum_{j=1}^{n} a_{ij}x_j \qquad [i = 1, \ldots, n] \qquad (7.2)$$

where the a_{ij} are the pseudo-first-order rate constants

$$a_{ij} = \frac{\partial f_i}{\partial x_j} \qquad [i,j = 1, \ldots, n] \qquad (7.3)$$

evaluated in the appropriate neighborhood of x_j. The assumption of linearity used to obtain Eqs. (7.2) from Eqs. (6.2) means only that the individual chemical steps are pseudo first order; they need not be unimolecular, and the sequence of reactions can still be cyclic or branched.

If the volume of the system is constant, the conservation of mass in a closed system requires that there exist positive numbers b_i such that

$$\sum_{i=1}^{n} b_i \dot{x}_i = \frac{d}{dt} \sum_{i=1}^{n} b_i x_i = 0 \qquad (7.4)$$

If the x_i are expressed as moles per unit volume, the b_i are the stoichiometric coefficients. It follows from Eqs. (7.2) and (7.4) that

$$\sum_{i=1}^{n} b_i \sum_{i=1}^{n} a_{ij} x_j = \sum_{j=1}^{n} \left(\sum_{i=1}^{n} \right) b_i a_{ij} \; x_j = 0 \qquad (7.5)$$

Since these relationships must be valid for all values of the x_j, the conservation of mass requires that the a_{ij} matrix be singular:

$$|a_{ij}| = 0 \qquad (7.6)$$

The linear dependence of the rows of Eq. (7.6) can be expressed in terms of positive constants:

$$\sum_{i=1}^{n} b_i a_{ij} = 0 \qquad [j = 1, \ldots, n] \qquad (7.7)$$

$$b_i > 0 \qquad [i = 1, \ldots, n] \qquad (7.8)$$

It is clear therefore that all the a_{ij} cannot be of the same sign. The requirement that chemical concentrations be nonnegative restricts the a_{ij} [i ≠ 1] to be nonnegative too:

$$a_{ij} \geq 0 \qquad [i \neq j;\ i, j = 1, \ldots, n] \qquad (7.9)$$

If any of these off-diagonal a_{ij} [i ≠ j] were negative, it would be possible for one or more of the chemical concentrations to be negative.

The requirement resulting from the conservation of mass [Eq. (7.6)] and the requirement implied by the fact that chemical concentrations cannot be negative [inequality (7.9)] lead to the conclusion that the diagonal elements of $|a_{ij}|$ must be nonpositive.

$$a_{ii} = \frac{1}{b_i} \sum_{j=1}^{n} b_j a_{ij} \leq 0 \qquad [i = 1, \ldots, n] \tag{7.10}$$

These results impose severe restrictions on the possible physical interpretations of Eqs. (7.2). For example, since an off-diagonal element in $|a_{ij}|$ cannot be negative, inhibition cannot be expressed by the linearized differential equations. Thus commonly encountered biochemical events such as end-product feedback inhibition and other forms of enzyme inhibition are inherently nonlinear and cannot be described by linearized differential equations. Evidently, the proper physical interpretation of Eqs. (7.2) is that the rate of change of the ith chemical component consists of up to $n - 1$ positive terms and one negative term. The negative term describes the decomposition or loss of the ith component at a rate proportional to the concentration of itself. The positive terms, if any, describe the formation of the ith component from one or more of the other x_j $[j \neq 1]$ at rates proportional to the concentrations of each precursor.

According to the conservation of mass,

$$\sum_{i=1}^{n} b_i x_i(t) = \sum_{i=1}^{n} b_i x_i(0) = M = \text{constant} \tag{7.11}$$

Since every $b_i > 0$, the maximum value possible for any x_i is less than M/b_i. Since chemical concentrations are necessarily nonnegative, the minimum value possible for any x_i is zero. Thus, if the total mass of the system is finite, every x_i must be physically bounded:

$$0 \leq x_i < \frac{M}{b_i} \qquad [i = 1, \ldots, n] \tag{7.12}$$

Thus, closed chemical systems cannot be totally unstable. According to Theorem IV this means that the real part of the characteristic exponents for Eqs. (7.2) must be nonpositive. Furthermore, it can be shown[30] that these equations cannot have purely imaginary characteristic exponents if they describe a closed system.

The requirement that the diagonal elements of $|a_{ij}|$ be nonpositive is sufficient to ensure that the characteristic exponents of Eqs. (7.2) will be nonpositive real numbers when n = 2 or whenever there are only 2n - 2 off-diagonal nonzero a_{ij} [i = j]. For the general case the principle of microscopic reversibility[31] must be invoked in order to prove that the characteristic exponents of Eqs. (7.2) are real numbers. In particular, the requirement for finite free-energy changes implies that the zero elements in $|a_{ij}|$ must occur symmetrically about the diagonal. These considerations lead us to Hearon's general theorem:

> Theorem VII (Hearon): *For closed chemical systems obeying pseudo-first-order rate laws and the principle of microscopic reversibility, the characteristic exponents* [*of Eqs. (7.2)*] *are real nonpositive numbers; the concentrations of the chemical components in such systems cannot be periodic functions of time.*

7.2 DEVELOPMENT OF DYNAMIC EXPRESSIONS FOR OPEN SEQUENTIAL MULTIENZYME SYSTEMS

Metabolic reactions involve sequences of individual enzyme-catalyzed reactions. Usually it is possible to write a nonnegative function $f_i^+(x_1, \ldots, x_n)$ that describes the rate of formation of the ith metabolite, and a nonnegative function, $f_i^-(x_1, \ldots, x_n)$ that describes the rate of disappearance of this compound. Then the differential equations for the system are

$$x_i = f_i^+(x_1, \ldots, x_n) - f_i^-(x_1, \ldots, x_n) \qquad [i = 1, \ldots, n] \tag{7.13}$$

If the metabolic pathway is a *linear* sequence of enzyme-catalyzed reactions, x_1 is converted to x_2 via enzyme E_1, x_i is converted to x_{i+1} via E_i, and x_{n-1} is converted to x_n via E_{n-1}:

$$X_1 \overset{E_1}{=} X_2 \cdots X_i \overset{E_i}{=} X_{i+1} \cdots X_{n-1} \overset{E_{n-1}}{=} X_n \tag{7.14}$$

If we assume that the interconversions between X_i and X_{i+1} are stiochiometric, and that the rate of the reaction catalyzed by E_i depends only on X_i and X_{i+1} (the substrate and the product of the enzyme), the differential equations are

$$\dot{x}_1 = f_1^+(x_2) - f_1^-(x_1)$$

$$\dot{x}_i = f_i^+(x_{i-1}, x_{i+1}) - f_i^-(x_i) \qquad [i = 2, \ldots, n - 1]$$

$$\dot{x}_n = f_n^+(x_{n-1}) - f_n^-(x_n) \tag{7.15}$$

In general the f's in Eqs. (7.15) are the ratios of polynomials in the respective arguments (cf. Chapters 1 and 2):

$$\dot{x}_1 = \frac{\sum_{j=1}^{p_{1r}} {}^{1r}\alpha_j x_2^{\,j}}{1 + \sum_{j=1}^{q_{1r}} {}^{1r}\beta_j x_2^{\,j}} - \frac{\sum_{j=1}^{p_{1f}} {}^{1f}\alpha_j x_1^{\,j}}{1 + \sum_{j=1}^{q_{1f}} {}^{1f}\beta_j x_1^{\,j}}$$

$$\dot{x}_i = \frac{\sum_{j=1}^{p_{i-1f}} {}^{i-1f}\alpha_j x_{i-1}^{\,j}}{1 + \sum_{j=1}^{q_{i-1f}} {}^{i-1f}\beta_j x_{i-1}^{\,j}} + \frac{\sum_{j=1}^{p_{ir}} {}^{ir}\alpha_j x_{i+1}^{\,j}}{1 + \sum_{j=1}^{q_{ir}} {}^{ir}\beta_j x_{i+1}^{\,j}}$$

$$- \left[\frac{\sum_{j=1}^{p_{if}} {}^{if}\alpha_j x_i^{\,j}}{1 + \sum_{j=1}^{q_{if}} {}^{if}\beta_j x_i^{\,j}} + \frac{\sum_{j=1}^{p_{i-1r}} {}^{i-1r}\alpha_j x_i^{\,j}}{1 + \sum_{j=1}^{q_{i-1r}} {}^{i-1r}\beta_j x_i^{\,j}} \right]$$

$[i = 2, \ldots, n - 1]$

$$\dot{x}_n = \frac{\sum_{j=1}^{p_{n-1f}} {}^{n-1f}\alpha_j x_{n-1}^j}{1 + \sum_{j=1}^{q_{n-1f}} {}^{n-1f}\beta_j x_{n-1}^j} - \frac{\sum_{j=1}^{p_{n-1r}} {}^{n-1r}\alpha_j x_n^j}{1 + \sum_{j=1}^{q_{n-1r}} {}^{n-1r}\beta_j x_n^j} \tag{7.16}$$

The individual rate functions in Eqs. (7.16) may have extrema, inflections, etc., or they may be rectangular hyperbolas. In the case of quasi-steady-state enzyme mechanisms (cf. Chapter 2) described by such hyperbolic rate functions, the differential equations are

$$\dot{x}_1 = \frac{{}^1V_r x_2}{{}^1K_r + x_2} - \frac{{}^1V_f x_1}{{}^1K_f + x_1}$$

$$\dot{x}_i = \frac{{}^{i-1}V_f x_{i-1}}{{}^{i-1}K_f + x_{i-1}} + \frac{{}^{i}V_r x_{i+1}}{{}^{i}K_f + x_{i+1}} - \left[\frac{{}^{i}V_f x_i}{{}^{i}K_f + x_i} + \frac{{}^{i-1}V_r x_i}{{}^{i-1}K_r + x_i}\right]$$

$$[i = 2, \ldots, n - 1]$$

$$\dot{x}_n = \frac{{}^{n-1}V_f x_{n-1}}{{}^{n-1}K_f + x_{n-1}} - \frac{{}^{n-1}V_r x_n}{{}^{n-1}K_r + x_n} \tag{7.17}$$

where the ${}^iV_f = {}^{if}\alpha_1/{}^{if}\beta_1$, ${}^iV_r = {}^{ir}\alpha_1/{}^{ir}\beta_1$, ${}^iK_f = 1/{}^{if}\beta_1$, and ${}^iK_r = 1/{}^{ir}\beta_1$.

Linearization of Eqs. (7.15) leads to

$$\dot{x}_1 = a_{12}x_2 + a_{11}x_1$$

$$\dot{x}_i = a_{i,i-1}x_{i-1} + a_{i,i+1}x_{i+1} + a_{i,i}x_i \qquad [i = 2, \ldots, n - 1]$$

$$\dot{x}_n = a_{n,n-1}x_{n-1} + a_{n,n}x_n \tag{7.18}$$

where $a_{11} = -\partial \bar{f}_1/\partial x_1$, $a_{ii} = -\partial \bar{f}_i/\partial x^i = -(a_{i-1,i} + a_{i+1,i})$ $[i = 2, \ldots, n - 1]$, and $a_{nn} = -\partial \bar{f}_n/\partial x_n$ are the negative entries on the $|a_{ij}|$ diagonal, and $a_{12} = \partial f_1^+/\partial x_2$, $a_{i,i-1} = \partial f_1^+/\partial x_{i-1}$ and $a_{i,i+1} = \partial f_i^+/\partial x_{i,i+1}$ $[i = 2, \ldots, n - 1]$, and $a_{n,n-1} = \partial f_n^+/\partial x_{n-1}$ are the positive off-diagonal elements. In each case the a_{ij} are calculated by evaluating the partial derivative at the value of x_i where the linearization is to be used. In the case of hyperbolic rate laws [Eqs. (7.17)],

$$f(x) = \frac{Vx}{K + x} \tag{7.19}$$

linearization in the neighborhood of $x = x^*$ leads to

$$a = \left.\frac{f}{x}\right|_{x=x^*} = \frac{VK}{(K + x^*)^2} \tag{7.20}$$

It is clear from Eq. (7.20) that the linearization procedure can be carried out for any value of x^*, and the resulting pseudo-first-order rate constant will be approximately valid for some small neighborhood of this concentration of x. In the special case $x^* << K$, the linearization procedure is valid when the substrate concentration is small compared to the Michaelis constant. Then

$$a = \frac{V}{K} \tag{7.21}$$

and the pseudo-first-order rate constant for an otherwise hyperbolic enzyme-catalyzed reaction is the ratio of the maximum rate to the Michaelis constant (cf. Chapter 1). However, for all larger values of x^*, $a < V/K$.

If the individual forward and reverse steps in Eqs. (7.14) are all pseudo first order, the linear differential equations describing the system are

$$\dot{x}_1 = K_{-1}x_2 - K_1x_1$$

$$\dot{x}_i = K_{i-1}x_{i-1} + K_{-i}x_{i+1} - (K_{-(i-1)} + K_i)x_i \qquad [i = 2, \ldots, n - 1]$$

$$\dot{x}_n = K_{n-1}x_{n-1} - K_{-(n-1)}x_n \tag{7.22}$$

Equations (7.22) illustrate that the solution at thermodynamic equilibrium, $K_{eq} \equiv K_i/K_{-i} = x_{i+1}(\infty)/x_i(\infty)$, is the only solution for which all the $\dot{x}_i = 0$. Thus, the closed chemical system in Eqs. (7.14) cannot possess a nonequilibrium stationary state. Furthermore, since *equilibrium* stationary states are asymptotically stable, the mechanism in Eqs. (7.14) is inherently stable too.

Of course, metabolic reactions are neither isolated nor closed systems. There is usually a flux of matter into metabolic sequences, and there may also be an exchange of energy with the "outside" world. Often there is a substance upstream in the metabolic sequence that is replenished by a large external reservior, or by the preceeding metabolic step. In what follows we arbitrarily designate this substance as x_0. If x_0 is converted to x_1, the differential equation describing x_1 is

$$\dot{x}_1 = F(x_0) + f_1^+(x_2) - f_1^-(x_1) \tag{7.22a}$$

where $F(x_0)$ describes the rate of introduction of x_1 from x_0, and $f_1^-(x_1)$ now includes the term describing the reverse of this step. If the conversion of x_0 to x_1 is catalyzed via enzyme E_0,

$$F(x_0) = \frac{\sum_{j=1}^{p_{0f}} {}^{0f}\alpha_j x_0^j}{1 + \sum_{j=1}^{q_{0f}} {}^{0f}\beta_j x_0^j} \tag{7.22b}$$

$$f_1^-(x_1) = -\left[\frac{\sum_{j=1}^{p_{1f}} {}^{1f}\alpha_j x_1^j}{1 + \sum_{j=1}^{q_{1f}} {}^{1f}\beta_j x_1^j} + \frac{\sum_{j=1}^{p_{0r}} {}^{0r}\alpha_j x_1^j}{1 + \sum_{j=1}^{q_{0r}} {}^{0r}\beta_j x_1^j}\right] \tag{7.22c}$$

If x_0 is maintained constant, $F(x_0) = K_0$ = constant, and the differential equation for x_1 is

$$\dot{x}_1 = K_0 + f_1^+(x_2) - f_1^+(x_1) \tag{7.22d}$$

Of course, x_0 need not really be the *first* upstream substance in a metabolic sequence; it can be any upstream metabolite whose concentration is maintained by its source.

There is also almost always a way for downstream metabolites to be removed either by diffusion or by a subsequent metabolic step. In what follows we arbitrarily designate this substance x_{n+1}. If x_n is converted to x_{n+1} via enzyme E_n, the differential equation for x_n is

$$\dot{x}_n = f_n^+(x_{n-1}, x_{n+1}) - f_n^-(x_n) \tag{7.22e}$$

where f_n^+ now includes terms describing the formation of x_n in the reverse reaction catalyzed by E_n,

$$f_n^+(x_{n-1}, x_{n+1}) = \frac{\sum_{j=1}^{p_{n-1f}} {}^{n-1f}\alpha_j x_{n-1}^j}{1 + \sum_{j=1}^{q_{n-1f}} {}^{n-1f}\beta_j x_{n-1}^j} + \frac{\sum_{j=1}^{p_{nf}} {}^{nf}\alpha_j x_{n+1}^j}{1 + \sum_{j=1}^{q_{nf}} {}^{nf}\beta_j x_{n+1}^j} \tag{7.22f}$$

and f_n^- includes the terms describing the conversion of x_n to x_{n+1}:

$$f_n^-(x_n) = \frac{\sum_{j=1}^{p_{n-1r}} {}^{n-1r}\alpha_j x_n^j}{1 + \sum_{j=1}^{q_{n-1r}} {}^{n-1r}\alpha_j x_n^j} + \frac{\sum_{j=1}^{p_{nf}} {}^{nf}\alpha_j x_n^j}{1 + \sum_{j=1}^{q_{nf}} {}^{nf}\alpha_j x_n^j} \tag{7.22g}$$

The differential equation for x_{n+1} is

$$\dot{x}_{n+1} = f_{n+1}^+(x_n, Y) - f_{n+1}^-(x_{n+1}) \tag{7.22h}$$

where Y is some parameter related to the reverse of the step wherein x_{n+1} is lost to the exterior world. f_{n+1}^+ is composed of two terms,

$$f^{+}_{n+1}(x_n, Y) = \frac{\sum_{j=1}^{p_{nf}} {}^{nf}\alpha_j x_n^j}{1 + \sum_{j=1}^{q_{nf}} {}^{nf}\beta_j x_n^j} + f^{+}(Y) \qquad (7.22i)$$

The first term on the right-hand side of Eq. (7.22i) describes the conversion of x_n to x_{n+1} via enzyme E_n. $f^{+}(Y) = K_{-(n+1)}$ = constant describes the reverse of the step wherein x_{n+1} is lost to the "outside" world. f^{-}_{n+1} is also composed of two terms,

$$f^{-}_{n+1}(x_{n+1}) = \frac{\sum_{j=1}^{p_{nr}} {}^{nr}\alpha_j x_{n+1}^j}{1 + \sum_{j=1}^{q_{nr}} {}^{nr}\beta_j x_{n+1}^j} + f^{-}(x_{n+1}) \qquad (7.22j)$$

The first term on the right-hand side of Eq. (7.22j) describes the reverse reaction catalyzed by E_n (i.e., the reaction wherein x_{n+1} is converted to x_n), and $f^{-}(x_{n+1})$ describes the rate of loss of x_{n+1} to the "exterior."

The complete model for the open metabolic system described by Eqs. (7.22) contains n + 1 variables:

$$= X_0 \;\Big|\; = X_0 \overset{E_0}{=} X_1 \overset{E_1}{=} X_2 \cdots X_i \overset{E_i}{=} X_{i+1} \cdots X_{n-1} \overset{E_{n-1}}{=} X_n \overset{E_n}{=} X_{n+1} \;\Big|\; = X_{n+1} = \qquad (7.23)$$

The complete differential equations are

$$\dot{x}_1 = K_0 + f_1^{+}(x_2) - f_1^{-}(x_1)$$

$$\dot{x}_i = f_i^{+}(x_{i-1}, x_{i+1}) - f_i^{-}(x_i) \qquad [i = 2, \ldots, n]$$

$$\dot{x}_{n+1} = f^{+}_{n+1}(x_n) + K_{-(n+1)} - f^{-}_{n+1}(x_{n+1}) \qquad (7.24)$$

If the enzyme-catalyzed steps are described by hyperbolic functions, these equations are

$$\dot{x}_1 = K_0 + \frac{{}^1V_r x_2}{{}^1K_r + x_2} - \frac{{}^1V_f x_1}{{}^1K_f + x_1} + \frac{{}^0V_r x_1}{{}^0K_r + x_1}$$

$$\dot{x}_i = \frac{{}^{i-1}V_f x_{i-1}}{{}^{i-1}K_f + x_{i-1}} + \frac{{}^iV_r x_{i+1}}{{}^iK_r + x_{i+1}} - \frac{{}^iV_f x_i}{{}^iK_f + x_i} + \frac{{}^{i-1}V_r x_i}{{}^{i-1}K_r + x_i}$$

$$[i = 2, \ldots, n]$$

$$\dot{x}_{n+1} = \frac{{}^nV_f x_n}{{}^nK_f + x_n} + K_{-(n+1)} - \frac{{}^nV_r x_{n+1}}{{}^nK_r + x_{n+1}} - f^-(x_{n+1}) \qquad (7.25)$$

If the individual steps in Eqs. (7.23) are all pseudo first order, the linearized differential equations are

$$\dot{x}_1 = K_0 + K_{-1}x_2 - K_1x_1 - K_{-0}x_1$$

$$\dot{x}_i = K_{i-1}x_{i-1} + K_{-i}x_{i+1} - \left[K_{-(i-1)} + K_i\right]x_i \qquad [i = 2, \ldots, n]$$

$$\dot{x}_{n+1} = K_n x_n + K_{-(n+1)} - (K_{-n} + K_{n+1})x_{n+1} \qquad (7.26)$$

Equations (7.24)-(7.26) illustrate that thermodynamic equilibrium is *not* the only solution for which all the $\dot{x}_i = 0$. Thus, the mechanism described in Eqs. (7.23) can possess nonequilibrium stationary states.

Since Eqs. (7.22) and (7.23) describe fully reversible chemical steps, there is an equilibrium solution at which the net rate for each step is zero (i.e., the forward reaction rate is exactly balanced by the rate of the reverse reaction). As usual, this equilibrium solution is asymptotically stable. Thus, the effect of including nonzero, finite rates for each of the reverse steps in Eqs. (7.23) is to enhance the asymptotic stability of the whole system.

On the other hand, if the model is a far-from-equilibrium

system, the steps in Eqs. (7.23) may be written as irreversible reactions:

$$\longrightarrow X_0 \; \Big| \longrightarrow X_0 \xrightarrow{E_0} X_1 \xrightarrow{E_1} X_2 \cdots X_i \xrightarrow{E_i} X_{i+1} \cdots X_{n-1} \xrightarrow{E_{n-1}} X_n \xrightarrow{E_n} X_{n+1} \Big| \longrightarrow X_{n+1} \longrightarrow \qquad (7.27)$$

The mechanism described by system (7.27) cannot possess a near-equilibrium solution. The differential equations are

$$\dot{x}_1 = K_0 - f_1(x_1)$$

$$\dot{x}_i = f_{i-1}(x_{i-1}) - f_i(x_i) \qquad [i = 2, \ldots, n]$$

$$\dot{x}_{n+1} = f_n(x_n) - f_{n+1}(x_{n+1}) \qquad (7.28)$$

where

$$f_i(x_i) = \frac{\sum_{j=1}^{p_i} {}^i\alpha_j x_i^j}{1 + \sum_{j=1}^{q_i} {}^i\beta_j x_i^j} \qquad [i = 1, \ldots, n + 1] \qquad (7.29)$$

and K_0 describes the constant, irreversible introduction of x_0 into the system, and $f_{n+1}(x_{n+1})$ describes the irreversible loss of x_{n+1} to the exterior.

If the enzyme-catalyzed steps are described by hyperbolas, the differential equations for the far-from-equilibrium system are

$$\dot{x}_1 = K_0 - \frac{{}^1Vx_1}{{}^1K + x_1}$$

$$\dot{x}_i = \frac{{}^{i-1}Vx_{i-1}}{{}^{i-1}K + x_{i-1}} - \frac{{}^iVx_i}{{}^iK + x_i} \qquad [i = 2, \ldots, n]$$

$$\dot{x}_{n+1} = \frac{{}^nVx_n}{{}^nK + x_n} - f_{n+1}(x_{n+1}) \tag{7.30}$$

If the individual steps are pseudo first order, the linearized differential equations are

$$\begin{aligned} \dot{x}_1 &= K_0 - K_1x_1 \\ \dot{x}_i &= K_{i-1}x_{i-1} - K_ix_i \qquad [i = 2, \ldots, n + 1] \end{aligned} \tag{7.31}$$

Far-from-equilibrium chemical reactions are common in metabolic pathways. The synthesis of complicated molecules and macromolecules, and the controlled utilization of energy in degradative pathways are both concrete examples of in vivo multienzyme systems that are not functioning near thermodynamic equilibrium. Furthermore, molecules for which cells have relatively constant requirements should be available at controlled, stationary levels, whereas molecules for which there are periodic requirements during the cell cycle may appear at peak levels and then disappear in oscillatory fashion. The extent to which these types of dynamic behavior can be achieved is related to the stability properties of the stationary states of the metabolites. It is thus important to obtain some idea about the expected dynamic behavior for the concentrations of substrates and products in far-from-equilibrium multienzyme systems.

7.3 FAR-FROM-EQUILIBRIUM STATIONARY STATES

Equations (7.30) admit the possibility of not more than one nonequilibrium stationary state. If such a state exists, it occurs when

$$x_i^* = \frac{{}^iK\, K_0}{{}^iV - K_0} \qquad [i = 1, \ldots, n]$$

$$f(x_{n+1}^*) = K_0 \tag{7.32}$$

Since chemical concentrations are finite and nonnegative,

$$I_V > K_0 \qquad [i = 1, \ldots, n] \tag{7.33}$$

is a requirement for the existance of *any* stationary state for Eqs. (7.30). In other words, for multienzyme systems with rates described by hyperbolic functions, it is necessary for the maximum velocity of each catalyzed step to exceed the rate of flux of x_0 into the pathway. If one of the maximum velocities is too small, the system will not possess a stationary state, and the intermediates will accumulate indefinitely.

Equations (7.31) on the other hand, suggest that there will always be one and only one nonequilibrium stationary state for linearized multienzyme systems. This stationary state arises when

$$x_i^* = \frac{K_0}{K_i} \qquad [i = 1, \ldots, n+1] \tag{7.34}$$

If we modify system (7.27) slightly and allow the individual enzyme-catalyzed steps to be reversible,

$$\longrightarrow X_0 \longrightarrow \boxed{X_0 \xrightarrow{E_0} X_1 \overset{E_1}{=} X_2 \cdots X_i \overset{E_i}{=} X_{i+1} \cdots X_{n-1} \overset{E_{n-1}}{=} X_n \overset{E_n}{=} X_{n+1}} \longrightarrow X_{n+1} \longrightarrow \tag{7.35}$$

the linearized differential equations are

$$\dot{x}_1 = K_0 + K_{-1}x_2 - K_1x_1$$

$$\dot{x}_i = K_{i-1}x_{i-1} + K_{-i}x_{i+1} - (K_i + K_{-(i-1)})x_i \qquad [i = 2, \ldots, n]$$

$$\dot{x}_{n+1} = K_nx_n - (K_{n+1} + K_{-n})x_{n+1} \tag{7.36}$$

Equation (7.36) also suggest that there will be one and only one nonequilibrium stationary state,

$$x_i^* = \frac{K_0}{K_i}\frac{K_{i+1} + K_{-i}}{K_{i+1}} \qquad [i = 1, \ldots, n]$$

$$x_{n+1}^* = \frac{K_0}{K_{n+1}} \tag{7.37}$$

Comparison of Eqs. (7.34) and (7.37) illustrates that the stationary-state level of x_{n+1}^* is the same regardless of whether the steps catalyzed by the E_i are reversible or irreversible, but the stationary-state levels of all the upstream x_i are higher in the reversible mechanism. If all the pseudo-first-order rate constants are approximately the same, the stationary-state levels of the upstream components will be about twice that expected for the far-from-equilibrium mechanism.

If we introduce the usual transformation of variables [cf. Eqs. (6.4), $u_i = x_i - x_i^*$], the transformed differential equations are

$$\dot{u}_1 = -K_1u_1 + K_{-1}u_2$$

$$\dot{u}_i = K_{i-1}u_{i-1} - [K_i + K_{-(i-1)}]u_i + K_{-i}u_{i+1} \qquad [i = 2, \ldots, n]$$

$$\dot{u}_{n+1} = K_nu_n - (K_{-n} + K_{n+1})u_{n+1} \tag{7.38}$$

Equations (7.38) have the same form as Eqs. (7.1) or (7.22), but the variables in Eqs. (7.38) can be negative, and there is no applicable mass-conservation equation comparable to Eq. (7.11).

Equations (7.38) suggest, however, that the stability properties of the nonequilibrium stationary state may be similar to the properties of the equilibrium stationary states.

The (n + 1)-by-(n + 1) coefficient matrix for Eqs. (7.38) contain exactly 2n off-diagonal elements symmetrically located on each side of the diagonal. Because rate constants are nonnegative real numbers, these off-diagonal elements are also nonnegative and real:

$$a_{12} = K_{-1}$$
$$a_{i,i-1} = K_{i-1} \qquad a_{i,i+1} = K_{-i} \qquad [i = 2, \ldots, n] \tag{7.39}$$
$$a_{n,n+1} = K_n$$

Furthermore, the n + 1 diagonal elements are necessarily nonpositive real numbers:

$$\begin{aligned} a_{11} &= -K_1 \\ a_{ii} &= -[K_i + K_{-(i-1)}] \qquad [i = 2, \ldots, n + 1] \end{aligned} \tag{7.40}$$

Since the off-diagonal elements are nonnegative real numbers located adjacent to and on each side of the diagonal, and since the diagonal elements are nonpositive real numbers whose absolute values cannot be smaller than any entry in the same column,

$$-a_{ii} \geq a_{ij} \qquad [j = i;\ j,\ i = 1, \ldots, n + 1] \tag{7.41}$$

the nonzero characteristic exponents for Eqs. (7.38) are necessarily negative and real.[28] This guarantees that Eqs. (7.38) and systems like systems (7.35) will be asymptotically stable (cf. Theorem III in Chapter 6).

If, instead of system (7.35), we consider the irreversible system, the transformed differential equations are

$$\begin{aligned} \dot{u}_1 &= -K_1 u_1 \\ \dot{u}_i &= K_{i-1} u_{i-1} - k_i u_i \qquad [i = 2, \ldots, n + 1] \end{aligned} \tag{7.42}$$

where $u_i = x_i - x_i^*$ for the variables in Eq. (7.31), and the x_i^* are defined in Eqs. (7.34). As before, the (n + 1)-by-(n + 1)-coefficient matrix contains exactly n + 1 diagonal elements, but in the case of Eqs. (7.42) there are only n off-diagonal elements. The diagonal elements are nonpositive real numbers

$$a_{ii} = -K_i \qquad [i = 1, \ldots, n + 1] \tag{7.43}$$

and the off-diagonal elements are nonnegative real numbers

$$a_{i,i-1} = K_{i-1} \qquad [i = 2, \ldots, n + 1] \tag{7.44}$$

because the rate constants are nonnegative and real. Furthermore, since the off-diagonal elements are located to the left of and adjacent to the diagonal, the off-diagonal entries do not appear in the characteristic equation for Eqs. (7.42):

$$\prod_{i=1}^{n+1} (K_i + \mu) = 0 \tag{7.45}$$

The characteristic exponents are

$$\mu_i = -K_i \qquad [i = 1, \ldots, n + 1] \tag{7.46}$$

Since the K_i are nonnegative and real, the nonzero characteristic exponents are negative real numbers. This result is sufficient to guarantee that the stationary state $x_i = x_i^*$ is asymptotically stable.

Clearly, the mechanism in system (7.27) is a special case of the more general mechanism in system (7.35). The two mechanisms become more and more similar as the K_{-i} become less and less significant in the dynamics. Far from equilibrium, where system (7.27) is applicable, the system is asymptotically stable because the characteristic exponents are negative real numbers. The effect of introducing reversibility to one or more of the steps in such a system is to make the characteristic exponents even smaller (larger negative numbers). This suggests that the introduction of reversibility into a metabolic system has the effect of increasing the stability (or, in the case of unstable systems, of decreasing or eleminating the instability). On the other hand, the effect of neglecting reversibility in a real system has the effect of increasing the potential for instability. This is consistent with the fact that at the other limit (near equilibrium) chemical systems are *always* asymptotically stable. Far-from-equilibrium models like system (7.27) are convenient simplifications of reversible models because the differential equations are less complex when the steps involving the reverse reactions are neglected. Furthermore, stability studies of far-from-equilibrium systems should provide *sufficient* conditions for stability of the corresponding reversible models and, in addition, *necessary* conditions for their instability.

7.4 TWO-COMPONENT SYSTEMS

Most biochemical pathways involve more than two steps. The philosophical basis for using a two-component system to describe a multicomponent system is the *hope* that only two steps are important as far as the dynamics of the whole pathway is concerned. However, since equilibrium is an asymptotic state, concepts like "near equilibrium" and "far from equilibrium" are necessarily relative. There is probably no single "pace maker" step that completely determines the dynamics for any metabolic sequence; in real situations *most* of the steps probably make *some* contribution to the overall dynamics and control. Two-component systems refer to pathways where all the *essential* features of the dynamics arise from the dynamics of two specific processes. Representations of multicomponent systems by two key components is clearly a dangerous pastime, since many features of systems involving two steps cannot be extrapolated to otherwise identical systems involving three steps.[32] The reader is therefore warned that results from systems involving two components may or may not apply in even a qualitative way to more complex systems.

Autonomous two-component chemical systems can be described by [cf. Eqs. (7.13)]

$$\dot{x}_1 = f_1^+(x_1, x_2) - f_1^-(x_1, x_2)$$
$$\dot{x}_2 = f_2^+(x_1, x_2) - f_2^-(x_1, x_2) \tag{7.47}$$

where f_1^+, f_1^-, f_2^+, and f_2^- are all positive functions of x_1 and x_2. For open systems, f_1^+ describes the rate of introduction of x_1 into the system and the rate of formation of x_1 within the system. f_1^- and f_2^- describe the rates at which x_1 and x_2 disappear, and f_2^+ describes the rate at which x_2 appears.

In enzyme-catalyzed systems, f_1^+ might depend on x_1 and/or x_2 if, for example, x_1 and/or x_2 affect the catalytic activity of the enzyme E_0 that converts x_0 to x_1. In reversible systems, f_1^+ is expected to depend on x_2 in a manner that takes into account the reverse reaction catalyzed by E_1 (wherein x_2 is converted to x_1).

f_1^- is expected to depend upon x_1 because x_1 is the substrate for E_1. f_1^- might also depend on x_2 if, for example, the activity of E_1 is altered by x_2. If all of the x_1 is converted stiochiometrically into x_2,

$$f_1^- = f_2^+ \tag{7.48}$$

but if some x_1 "leaks" out of the system,

$$f_1^- > f_2^+ \tag{7.49}$$

In either case, f_2^+ depends on x_1 because x_1 is the precursor of x_2, and f_2^+ can depend on x_2 if, for example, x_2 effects E_1.

Similarly, f_2^- depends on x_2, and it may also depend on x_1 if, for example, x_1 effects the process by which x_2 disappears.

The f's in Eq. (7.47) may be monotonic functions of x_1 and x_2, or they may experience extrema. If f is a monotonically decreasing function of a variable,

$$\frac{\partial f}{\partial x} \leq 0 \tag{7.50}$$

we say that the step described by f is *inhibited* by the variable. If, on the other hand, f is a monotonically increasing function of a variable,

$$\frac{\partial f}{\partial x} \geq 0 \tag{7.51}$$

we say that the step is *stimulated* by x.

In biological systems, substances that stimulate a step in one concentration range may inhibit the same step in a different concentration range. For example, at low concentrations, substrates obviously stimulate the activity of their enzymes, but at higher concentrations, some substrates inhibit their enzymes. For substrates that do not inhibit,

$$\frac{\partial f_1^-}{\partial x_1} \geq 0 \qquad \frac{\partial f_2^+}{\partial x_1} \geq 0 \qquad \frac{\partial f_2^-}{\partial x_2} \geq 0 \tag{7.52}$$

The sum of the partial derivatives for Eqs. (7.47) is

$$\frac{\partial \dot{x}_1}{\partial x_1} + \frac{\partial \dot{x}_2}{\partial x_2} = \frac{\partial f_1^+}{\partial x_1} + \frac{\partial f_2^+}{\partial x_2} - \frac{\partial f_1^-}{\partial x_1} - \frac{\partial f_2^-}{\partial x_2} \tag{7.53}$$

For substrates that do not inhibit their enzymes, the right-hand-side of Eq. (7.53) is nonpositive provided $\partial f_1^+/\partial x_1$ and $\partial f_2^+/\partial x_2$ are not positive. These conditions are obtained whenever (1) f_1^+ is independent of or inhibited by x_1, and (2) f_2^+ is independent of or inhibited by x_2. Under these conditions the left-hand-side of Eq. (7.53) does not change sign in the phase space, and according to Bendixson's[29] negative criterion (Theorem VI), Eqs. (7.47) cannot exhibit limit-cycle behavior. This result would not apply if more than two variables were required to describe the dynamics.[32]

If the step described by f_1^+ is stimulated by x_1, or the step described by f_2^+ is stimulated by x_2,

$$\frac{\partial f_1^+}{\partial x_1} \geq 0 \qquad \frac{\partial f_2^+}{\partial x_2} \geq 0 \tag{7.54}$$

Under these conditions, the left-hand side of Eq. (7.53) changes sign in the phase space iff

$$\frac{\partial f_1^{+}}{\partial x_1} + \frac{\partial f_2^{+}}{\partial x_2} = \frac{\partial f_1^{-}}{\partial x_1} + \frac{\partial f_2^{-}}{\partial x_2} \tag{7.55}$$

Furthermore, for substrates that do not inhibit their enzymes, $\partial f_1^{-}/\partial x_1$ or $\partial f_2^{-}/\partial x_2$ can individually change sign. Under these conditions, the left-hand side of Eq. (7.53) can change sign in the phase space, and the possibility of limit-cycle behavior cannot be ruled out by Theorem VI. These results are summarized in Theorem VIII:

> Theorem VIII: *Limit-cycle behavior in chemical systems described by Eqs. (7.47) is not possible if the components do not stimulate their own formation or inhibit their own production. This result is applicable only to two-component systems.*

If the model in system (7.23) involves only two components, the mechanism is

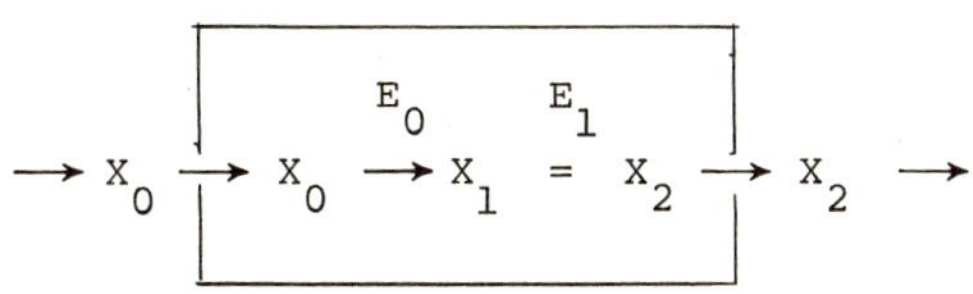

(7.56)

and the differential equations are

$$\begin{aligned} \dot{x}_1 &= f_1^{+}(x_2) - f_1^{-}(x_1) \\ \dot{x}_2 &= f_2^{+}(x_1) - f_2^{-}(x_2) \end{aligned} \tag{7.57}$$

The sum of the partial derivatives in Eq. (7.53) is

$$\frac{\partial x_1}{\partial x_1} + \frac{\partial x_2}{\partial x_2} = -(K_1 + K_2) < 0 \tag{7.58}$$

On the other hand, if the steps are hyperbolic, the differential equations are

$$\dot{x}_1 = K_0 + \frac{{}^1V_r x_2}{{}^1K_r + x_2} - \frac{{}^1V_f x_1}{{}^1K_f + x_1}$$
$$\dot{x}_2 = \frac{{}^1V_f x_1}{{}^1K_f + x_1} - \frac{{}^2V_f x_2}{{}^2K_f + x_2} \tag{7.59}$$

and the sum of the partials is

$$\frac{\partial \dot{x}_1}{\partial x_1} + \frac{\partial \dot{x}_2}{\partial x_2} = -\left[\frac{{}^1V_f\,{}^1K_f}{({}^1K_f + x_1)^2} + \frac{{}^2V_f\,{}^2K_f}{({}^2K_f + x_2)^2}\right] < 0 \tag{7.60}$$

Equations (7.58) and (7.60) illustrate that no limit cycles can exist for either of the type systems described by the differential equations (7.57) and (7.59). However, in the general case when the steps are described by the ratio of polynomials, the differential equations are

$$\dot{x}_1 = K_0 + \frac{\sum_{j=1}^{p_{1r}} {}^{1r}\alpha_j x_2^j}{1 + \sum_{j=1}^{q_{1r}} {}^{1r}\beta_j x_2^j} - \frac{\sum_{j=1}^{p_{1f}} {}^{1f}\alpha_j x_1^j}{1 + \sum_{j=1}^{q_{1f}} {}^{1f}\beta_j x_i^j}$$
$$\dot{x}_2 = \frac{\sum_{j=1}^{p_{1f}} {}^{1f}\alpha_j x_1^j}{1 + \sum_{j=1}^{q_{1f}} {}^{1f}\beta_j x_1^j} - \frac{\sum_{j=1}^{p_{2f}} {}^{2f}\alpha_j x_2^j}{1 + \sum_{j=1}^{q_{2f}} {}^{2f}\beta_j x_2^j} \tag{7.61}$$

and the Bendixson's[29] negative criterion is satisfied provided[33]

$$\frac{{}^{1f}\alpha_i}{{}^{1f}\beta_i} > \frac{{}^{1f}\alpha_{i-1}}{{}^{1f}\beta_{i-1}} \qquad [i = 2, \ldots, n] \tag{7.62}$$

$$\frac{{}^{2f}\alpha_i}{{}^{2f}\beta_i} > \frac{{}^{2f}\alpha_{i-1}}{{}^{2f}\beta_{i-1}} \qquad [i = 2, \ldots, n] \tag{7.63}$$

The inequalities in Eq. (7.62) are satisfied if

$$\frac{\sum_{j=1}^{p_{1f}} {}^{1f}\alpha_j x_1^{\,j}}{1 + \sum_{j=1}^{q_{1f}} {}^{1f}\beta_j x_1^{\,j}} \quad \text{versus} \quad x_1$$

does not experience an extremum, and the inequalities in (7.63) are satisfied if

$$\frac{\sum_{j=1}^{p_{2f}} {}^{2f}\alpha_j x_2^{\,j}}{1 + \sum_{j=1}^{q_{2f}} {}^{2f}\beta_j x_2^{\,j}} \quad \text{versus} \quad x_2$$

does not experience an extremum.

7.5 THE SENSITIVITY OF METABOLIC LEVEL TO PARAMETER VARIATIONS

A system is well controlled if the stationary state possesses global asymptotic stability and if, in addition, the stationary-state level of the variables is relatively insensitive to variations in the parameters of the system. The global nature of the stability is necessary only if the variables are not constrained to a well-defined neighborhood of the stationary state, and the asymptotic nature of the stability is necessary if oscillations might result in excursions of the variables byyond the desired limits. But regardless of the nature of the stability, a well-controlled system must be stable, and it should be insensitive to parameter variation.

According to the linearized differential equations describing the far-from-equilibrium multienzyme systems, the stationary-state levels of the x_i are given by the ratios of the flux of x_0 into the system K_0 divided by the pseudo-first-order rate constant for the conversion of x_i into x_{i+1} (K_i) [cf. Eq. (7.34)]. The sensitivity of the x_i^* to changes in K_0 is given by

$$\frac{\partial x_i^*}{\partial K_0} = \frac{1}{K_i} \qquad [i = 1, \ldots, n + 1] \tag{7.64}$$

Equation (7.64) illustrates that the change in x_i^* that results from an increment in the flux is positive and inversely proportional to K_i.

The sensitivity of x_i^* to changes in K_i is

$$\frac{\partial x_i^*}{\partial K_i} = -\frac{K_0}{(K_i)^2} \qquad [i = 1, \ldots, n + 1] \tag{7.65}$$

Equation (7.65) illustrates that the change in x_i^* that results from an increment in the pseudo-first-order constant for its conversion to x_{i+1} (or, in the case of x_{n+1}, for its loss to the exterior) is negative, proportional to K_0, and inversely proportional to the pseudo-first-order constant squared. Since x_i^* in Eq. (7.31) does not depend on any K_j [$j \neq i$],

$$\frac{\partial x_i^*}{\partial K_j} = 0 \qquad [i \neq j;\ i = 1, \ldots, n + 1] \tag{7.66}$$

Equation (7.66) illustrates that x_i^* is insensitive to changes in all the other pseudo-first-order constants.

Equations (7.33) describe the stationary-state levels of the intermediates in a linearized reversible sequence of enzyme-catalyzed reactions. The sensitivity to changes in K_0 for such systems is

$$\begin{aligned} \frac{\partial x_i^*}{\partial K_0} &= \frac{K_{i+1} + K_{-i}}{K_i K_{i+1}} \qquad [i = 1, \ldots, n] \\ \frac{\partial x_{n+1}^*}{\partial K_0} &= \frac{1}{K_{n+1}} \end{aligned} \tag{7.67}$$

Equations (7.64) and (7.67) illustrate that x^*_{n+1} is equally sensitive to flux changes in reversible and irreversible mechanisms. However, since

$$\frac{K_{i+1} + K_{i}}{K_{i+1}} \geq 1 \tag{7.68}$$

x^*_i [i = 1, ..., n] are more sensitive to changes in K_0 in the reversible mechanism. In the following pages we shall find more examples of systems that possess added stability apparently at the expense of parameter insensivity.

The sensivity of x^*_i to changes in K_i is

$$\begin{aligned} \frac{\partial x^*_i}{\partial K_i} &= -\frac{K_0}{(K_i)^2}\frac{K_{i+1} + K_{-i}}{K_{i+1}} \qquad [i = 1, \ldots, n] \\ \frac{\partial x^*_{n+1}}{\partial K_{n+1}} &= -\frac{K_0}{(K_{n+1})^2} \end{aligned} \tag{7.69}$$

Again x^*_{n+1} is equally sensitive in the reversible and irreversible models, but the other x^*_i are more sensitive in the reversible mechanism.

In the case of the reversible mechanism, the x^*_i [i = 1, ..., n] are also sensitive to K_{i+1} and K_{-i}:

$$\frac{\partial x^*_i}{\partial K_{i+1}} = -\frac{K_0}{(K_{i+1})^2}\frac{K_{-i}}{K_i} \qquad [i = 1, \ldots, n] \tag{7.70}$$

$$\frac{\partial x^*_i}{\partial K_{-i}} = \frac{K_0}{K_i K_{i+1}} \qquad [i = 1, \ldots, n] \tag{7.71}$$

An increment in K_{i+1} results in a decrease in x^*_i [Eq. (7.70)], and an increment in K_{-i} results in an increse in x^*_i [Eq. (7.71)].

If the various steps in the far-from-equilibrium model are described by hyperbolic functions, the stationary-state levels of the x_i are given by Eqs. (7.28). The sensitivity of x_i^* to changes in flux is

$$\frac{\partial x_i^*}{\partial K_0} = \frac{{}^iK_f\,{}^iV_f}{({}^iV_f - K_0)^2} \qquad [{}^iV_f > K_0,\ i = 1, \ldots, n + 1] \tag{7.72}$$

Equation (7.72) illustrates that an increase in K_0 causes an increase in x_i^*; this increase is proportional to the Michaelis constant for the enzyme that converts x_i to x_{i+1}. The sensivity also depends on ${}^iV_f$ and K_0: as ${}^iV_f$ approaches the flux, this sensivity becomes very large, but if ${}^iV_f >> K_0$, the sensivity tends to become inversely proportional to ${}^iV_f$.

The sensivity of x_i^* to changes in the Michaelis constant is

$$\frac{\partial x_i^*}{\partial\,{}^iK_f} = \frac{{}^iV_f}{{}^iV_f - K_0} \qquad [{}^iV_f > K_0,\ i = 1, \ldots, n + 1] \tag{7.73}$$

Equation (7.73) illustrates that a unit increase in the ith Michaelis constant will cause at least a unit increase in x_i^*. As ${}^iV_f$ approaches K_0, x_i^* becomes extremely sensitive to changes in its Michaelis constant.

The sensivity of x_i^* to changes in its maximum velocity is

$$\frac{\partial x_i^*}{\partial\,{}^iV_f} = -\frac{{}^iK_f K_0}{({}^iV_f - K_0)^2} \qquad [{}^iV_f > K_0,\ i = 1, \ldots, n + 1] \tag{7.74}$$

Equation (7.74) illustrates that an increase in ${}^iV_f$ causes a decrease in x_i^*. This decrease is proportional to the Michaelis constant, and it also depends upon ${}^iV_f$ and K_0: as ${}^iV_f$ approaches the flux, the sensivity becomes large, but if ${}^iV_f >> K_0$, the sensivity has as its limit $-{}^iK_fK_0/{}^iV_f^{\,2}$.

CHAPTER 8

CHEMICAL KINETICS OF MULTICOMPONENT SYSTEMS WITH FEEDBACK CONTROL

8.1 DEVELOPMENT OF DYNAMIC EXPRESSION FOR OPEN SEQUENTIAL MULTIENZYME SYSTEMS WITH FEEDBACK CONTROL

Recent discoveries about the detailed chemical operation of metabolic control mechanisms provide a sound basis for studying the stability properties of complex multienzyme systems. Perhaps the most fundamental of these discoveries is that highly specific control mechanisms operate through feedback devices to regulate the activity of enzymes. The establishment of closed loops in the biochemical orgainzation of cells provides the detail necessary for an understanding of metabolic control mechanisms and for studies of their stability properties.

Metabolites bearing no stereochemical relationship to the substrate or the product can alter the activity of certain enzymes. Often it is observed that enzymes affected in this manner synthesize substances that are precursors of the interacting metabolites. This situation is indicated in Eqs. (8.1):

$$X_0 = X_0 \overset{E_0}{=} X_1 \overset{E_1}{=} X_2 \cdots X_i \overset{E_i}{=} X_{i+1} \cdots X_{n-1} \overset{E_{n-1}}{=} X_n \overset{E_n}{=} X_{n+1} = X_{n+1}$$

$$X_{n+1} + E_0 \overset{\alpha_1}{=} E_0(X_{n+1})$$

$$X_{n+1} + E_0 - X_{n+1} \overset{\alpha_2}{=} E_0(X_{n+1})_2$$

$$\vdots \qquad\qquad \vdots$$

$$X_{n+1} + E_0(X_{n+1})_{\rho-1} \overset{\alpha_\rho}{=} E_0(X_{n+1})_\rho$$

(8.1)

If E_0 is the most active form of this enzyme, the situation is usually referred to as *end-product inhibition*. In that case, E_0 is inhibited at all concentrations of x_{n+1}, and the controlled flux through the step catalyzed by E_0 is

$$F(x_{n+1}) = \frac{K_0}{1 + \sum_{i=1}^{\rho} \alpha_i (x_{n+1})^i} \tag{8.2}$$

where K_0 is the uninhibited flux, and the α's are association constants for x_{n+1} and the various forms of E_0, and ρ is the maximum number of molecules of x_{n+1} that can associate with a molecule of E_0. If we can treat the individual equilibria between x_{n+1} and the various forms of E_0 as a concerted overall reaction wherein ρ molecules of x_{n+1} react with one molecule of E_0,

$$\rho X_{n+1} + E_0 \overset{\alpha}{=} E(X_{n+1})_\rho \tag{8.3}$$

the controlled flux can be described by

$$F(x_{n+1}) = \frac{K_0}{1 + (x_{n+1})^\rho} \tag{8.4}$$

where α is the macroscopic association constant for ρ molecules of x_{n+1} and one molecule of E_0, and ρ is the stoichiometry of the feedback inhibition by the end-product x_{n+1}.

Figure 8.1 illustrates that the controlled fluxes described by Eqs. (8.2) or (8.4) are monotonically decreasing functions of the end-product concentration. Since, for all concentrations of x_{n+1}, $\partial F/\partial x_{n+1} \leq 0$, the effect exerted by x_{n+1} is usually referred to as *negative feedback*. Negative feedback is inherently nonlinear; like other forms of inhibition, it cannot be represented in linearized equations without compromising the nonnegativity of chemical concentrations.[30]

Many examples of end-product inhibition of enzymes have been documented. In certain organisms or tissues, each of the amino acids alanine, arginine, cysteine, cystine, glycine, histidine, isoleucine, leucine, lysine, proline, serine, threonine, tryptophane, and valine is known to inhibit at least one of the enzymes involved in its synthesis. Another example of an enxyme subject to end-product inhibition is glutamine synthetase, which is inhibited by at least eight different end products.[34] Many other enzymes, including enzymes involved in the early steps of fatty-acid syn-

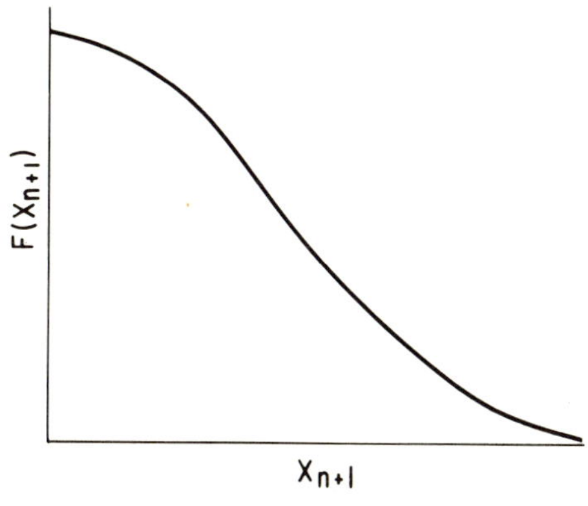

FIG. 8.1. A negative-feedback function.

thesis or nucleotide formation, are also subject to negative-feedback control by a substance formed toward the end of their metabolic pathways.

If E_0 in Eqs. (8.1) is the least active form of this enzyme, the situation is usually referred to as *end-product stimulation.* Enzyme stimulation by an immediate or distant product in the pathway occurs if the product relieves inhibition of E_0 by another compound, or if the product acts directly to increase the catalytic efficiency of E_0. In either case, if E_0 is stimulated at all concentrations of x_{n+1} by increasing x_{n+1}, the controlled flux through the step catalyzed by E_0 is

$$F(x_{n+1}) = K_0 \frac{1 + \sum_{i=1}^{\rho} \beta_i (x_{n+1})^i}{1 + \sum_{i=1}^{\rho} \alpha_1 (x_{n+1})^i} \qquad (8.5)$$

where $\beta_i > \alpha_i$ $[i = 1, \ldots, \rho]$. As before, if we can treat the individual equilibria between x_{n+1} and the various forms of E_0 as a concerted reaction, the controlled flux can be described by

$$F(x_{n+1}) = K_0 \frac{1 + \beta(x_{n+1})^\rho}{1 + \alpha(x_{n+1})^\rho} \qquad (8.6)$$

where $\beta > \alpha$, and the stimulated activity of $E_0(X_{n+1})_\rho$ is $K_0\beta/\alpha$.

Figure 8.2 illustrates that case when the controlled fluxes described by Eqs. (8.5) or (8.6) are monotonically increasing functions of the product concentration. Since, for all concentrations of x_{n+1}, $\partial F/\partial x_{n+1} \geq 0$, the effect exerted by x_{n+1} is referred to as *positive feedback.*

Several examples of product stimulation of enzymes have been documented, but in most cases it turns out that it is the immediate product that causes the increase in catalytic activity. The best-known example of product stimulation is the ADP stimulation of

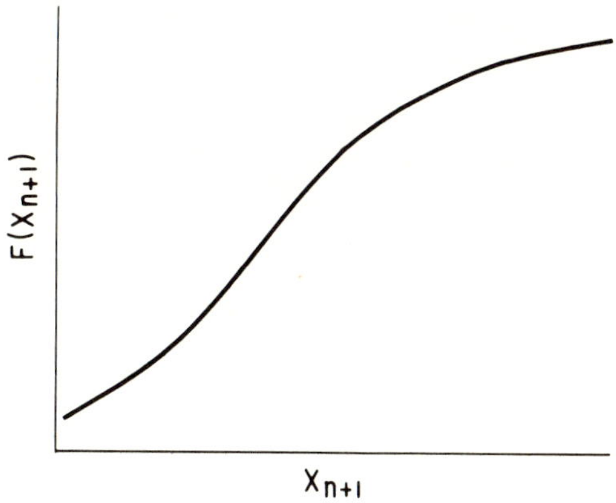

Fig. 8.2. A positive-feedback function.

phosphofructokinase (PFK) in yeast. ADP is both an immediate product of PFK,

$$\text{ATP} + \text{Fructose-6-phosphate} \xrightarrow{\text{PFK}} \text{ADP} + \text{Fructose-1,6-diphosphate}$$

and a distant product formed in the downstream step catalyzed by adenylate kinase (AK):

$$\text{AMP} + \text{ATP} \overset{\text{AK}}{=} 2\text{ADP}$$

End-product inhibition offers an efficient means of controlling metabolic fluxes when reactant concentrations are above normal, but this type of control device does not provide for economy of substances when metabolic levels are low. A combination of end-product inhibition by high metabolite concentrations and end-product stimulation at lower concentrations would provide a basis for the more stringent type of control necessary under these conditions. In this type of control mechanism, the enzyme site at which binding causes stimulation becomes saturated as metabolites approach their normal range. If the flux into the metabolic sequence is decreased, the metabolites fall below their normal range, and the activity of the controlling enzymes decline slightly. In this way, the metabolites are conserved when they are in short supply. However, if the flux into the metabolic sequence is increased so that the metabolic levels increase above their normal range, the activity of the controlling enzymes again decrease. In this way, the metabolic production is attinuated when there is excess supply.

Figure 8.3 illustrates the case when the controlled fluxes described by Eq. (8.6) experience an optimum at an intermediate ("normal") product concentration. Since, in this case, $\partial F/\partial x_{n+1}$ changes sign, the effect exerted by x_{n+1} includes both positive and negative feedback.

We expect the differential equations describing open multienzyme systems with feedback to be identical to the equations describing such systems without feedback [Eqs. (7.22)] except that K_0 should be replaced by $F(x_{n+1})$. For controlled, far-from-equilibrium systems,

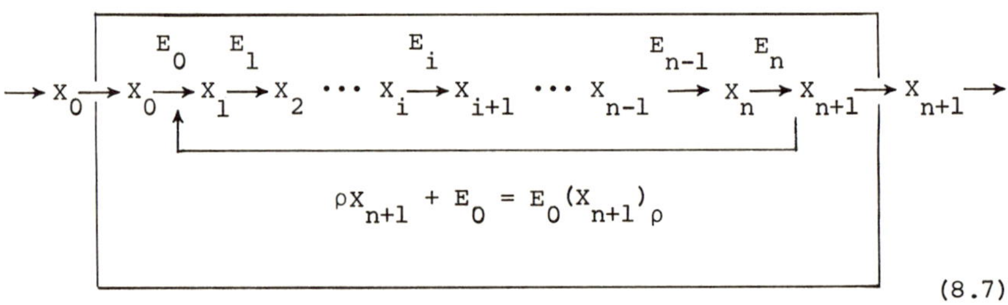

(8.7)

the differential equations are

$$\dot{x}_1 = F(x_{n+1}) - f_1(x_1)$$
$$\dot{x}_i = f_{i-1}(x_{i-1}) - f_i(x_i) \qquad [i = 2, \ldots, n + 1] \tag{8.8}$$

where the f_i are the usual functions describing the catalyzed steps, and F is the nonlinear function describing the feedback control by x_{n+1} on the step catalyzed by E_0.

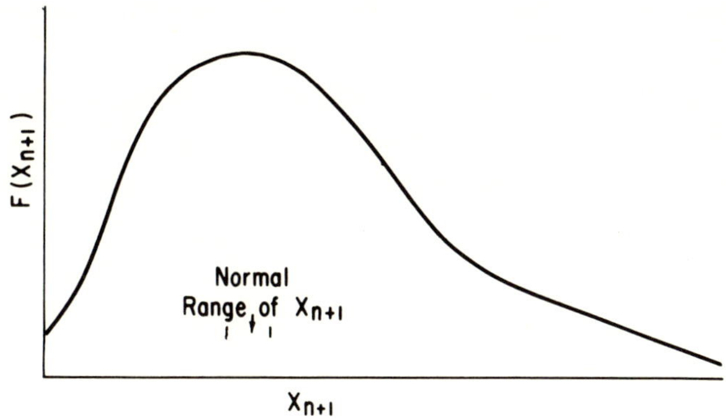

FIG. 8.3. A feedback function with positive feedback at low x_{n+1} and negative feedback at high x_{n+1}.

8.2 STATIONARY STATES FOR SYSTEMS WITH NEGATIVE FEEDBACK

The differential equations describing negative-feedback systems with enzyme-catalyzed steps described by hyperbolic functions are

$$\dot{x}_1 = \frac{K_0}{1 + \sum_{j=1}^{\rho} \alpha_j (x_{n+1})^j} - \frac{{}^1V x_1}{{}^1K + x_1}$$

$$\dot{x}_i = \frac{{}^{i-1}V x_{i-1}}{{}^{i-1}K + x_{i-1}} - \frac{{}^iV x_i}{{}^iK + x_i} \qquad [i = 2, \ldots, n + 1] \tag{8.9}$$

Since the feedback is negative, all the α's must be nonnegative, and the stationary-state solution,

$$\frac{K_0}{1 + \sum_{j=1}^{\rho} \alpha_j (x^*_{n+1})^j} - \frac{{}^1V x^*_i}{{}^1K + x^*_i} = 0$$

$$\frac{{}^{i-1}Vx^*_{i-1}}{{}^{i-1}K + x^*_{i-1}} - \frac{{}^{i}Vx^*_{i}}{{}^{i}K + x^*_{i}} = 0 \qquad [i = 2, \ldots, n + 1] \tag{8.10}$$

must have one real root for which $x^*_i > 0$, but according to the Descartes Rule, there cannot be more than one such root, because

$$^{n+1}KK_0 + (K_0 - {}^{n+1}V)x^*_{n+1} - {}^{n+1}V \sum_{j=1}^{\rho} \alpha_j (x^*_{n+1})^{j+1} = 0 \qquad [K_0 > {}^{n+1}V]$$

and

$$K_0 - K_{n+1}x^*_{n+1} + \left[K_{n+1} \sum_{j=1}^{\rho} \alpha_j (x^*_{n+1})^{j+1}\right] = 0$$

can alternate in sign once and only once. If $\rho \leq 3$, Eqs. (8.10) can be solved for the stationary state concentrations, x^* in radicals.[35] On the other hand, if $\rho > 3$, Eqs. (8.10) cannot be solved in radicals[36], and x^* can be obtained only by numerical methods.

The differential equations for negative feedback systems with pseudo-first-order enzyme-catalyzed steps are

$$\dot{x}_1 = \frac{K_0}{1 + \sum_{j=1}^{\rho} \alpha_j (x^*_{n+1})^j} - K_1x_1 \tag{8.11}$$

$$\dot{x}_i = K_{i-1}x_{i-1} - K_ix_i \qquad [i = 2, \ldots, n + 1]$$

Again, the stationary-state solution

$$\frac{K_0}{1 + \sum_{j=1}^{\rho} \alpha_j (x^*_{n+1})^j} - K_1x^*_1 = 0 \tag{8.12}$$

$$K_{i-1}x^*_{i-1} - K_ix^*_i = 0$$

must have one and only one real root for which $x^*_i > 0$, As before, the stationary-state solution cannot be obtained in radicals if $\rho > 3$.

8.3 STATIONARY STATES FOR SYSTEMS WITH POSITIVE FEEDBACK

The differential equations describing positive-feedback systems with enzyme-catalyzed steps described by hyperbolic functions are

$$\dot{x}_1 = K_0 \left[\frac{1 + \sum_{j=1}^{\rho} \beta_j (x_{n+1})^j}{1 + \sum_{j=1}^{\rho} \alpha_j (x_{n+1})^j} \right] - \frac{{}^1Vx_1}{{}^1K + x_1}$$

$$\dot{x}_i = \frac{{}^{i-1}Vx_{i-1}}{{}^{i-1}K + x_{i-1}} - \frac{{}^{i}Vx_i}{{}^{i}K + x_i} \qquad [i = 2, \ldots, n + 1] \tag{8.13}$$

In this case, the stationary-state solution

$$K_0 \left[\frac{1 + \sum_{j=1}^{\rho} \beta_j (x^*_{n+1})^j}{1 + \sum_{j=1}^{\rho} \alpha_j (x^*_{n+1})^j} \right] - \frac{{}^1Vx^*_1}{{}^1K + x^*_1} = 0$$

$$\frac{{}^{i-1}Vx^*_{i-1}}{{}^{i-1}K + x^*_{i-1}} - \frac{{}^{i}Vx^*_i}{{}^{i}K + x^*_i} = 0 \qquad [i = 2, \ldots, n + 1] \tag{8.14}$$

can possess more than one real root for which $x^*_i > 0$. Similarly, positive-feedback systems iwth pseudo-first-order catalytic steps are described by

$$\dot{x}_1 = K_0 \left[\frac{1 + \sum_{j=1}^{\rho} \beta_j (x_{n+1})^j}{1 + \sum_{j=1}^{\rho} \alpha_j (x_{n+1})^j} \right] - K_1 x_1$$

$$\dot{x}_i = K_{i-1}x_{i-1} - K_i x_i \qquad [i = 2, \ldots, n + 1] \tag{8.15}$$

The stationary-state solution

$$K_0\left[\frac{1+\sum_{j=1}^{\rho}\beta_j(x^*_{n+1})^j}{1+\sum_{j=1}^{\rho}\alpha_j(x^*_{n+1})^j}\right] - K_1x^*_1 = 0 \tag{8.16}$$

$$K_{i-1}x^*_{i-1} - K_ix^*_i = 0 \qquad [i = 2, \ldots, n + 1]$$

has only one real solution for which $x^*_i > 0$ when $\rho = 1$, because

$$K_0(K_0\beta_1 - K_{n+1})x^*_{n+1} - K_{n+1}\alpha_1(x^*_{n+1})^2 = 0$$

can alternate in sign once and only once. However, if ρ is odd and greater than unity, there could be as many as ρ real solutions for which $x^*_i > 0$; when ρ is even, there could be as many as $\rho + 1$ real $x^*_i > 0$.[37]

The possibility of multiple real and positive x^*_i for metabolic systems with positive feedback suggests that multiple stationary states could arise in such systems. If more than one of these stationary solutions possesses asymptotic stability, then it should be possible for real metabolic systems described by Eqs. (8.13) or (8.15) to exhibit more than one stable state and to switch between them.

8.4 LINEAR ANALYSIS OF BIOCHEMICAL SYSTEMS WITH FEEDBACK CONTROL

If the catalyzed steps are pseudo first order, Eqs. (8.8) become

$$\dot{x}_1 = F(x_{n+1}) - K_1x_1 \tag{8.17}$$

$$\dot{x}_i = K_{i-1}x_{i-1} - K_ix_i \qquad [i = 2, \ldots, n + 1]$$

The usual transformation of variables [cf. Eqs. (6.4), $u_i = x_i - x^*_i$] leads to the transformed differential equations

$$\dot{u}_1 = G(u_{n+1}) - K_1 u_1$$
$$\dot{u}_i = K_{i-1} u_{i-1} - K_i u_i \qquad [i = 2, \ldots, n+1] \tag{8.18}$$

where

$$G(u_{n+1}) = F(u_{n+1} + x^*_{n+1}) - K_1 x^*_1 \tag{8.19}$$

For negative-feedback systems wherein F is described by Eq. (8.2),

$$G(u_{n+1}) = \frac{K_0}{1 + \sum_{j=1}^{\rho} \alpha_j (u_{n+1} + x^*_{n+1})^j} - K_1 x^*_1 \tag{8.19a}$$

and for positive-feedback systems with F described by Eq. (8.5),

$$G(u_{n+1}) = K_0 \left[\frac{1 + \sum_{j=1}^{\rho} \beta_j (u_{n+1}) + x^*_{n+1})^j}{1 + \sum_{j=1}^{\rho} \alpha_j (u_{n+1} + x^*_{n+1})^j} \right] - K_1 x^*_1 \tag{8.19b}$$

In either case, the linearization of G in Eqs. (8.18) in the neighborhood of the stationary state $\dot{u}_i = 0$ [i = 1, ..., n + 1] leads to the abridged differential equations

$$\dot{u}_1 = A u_{n+1} - K_1 u_1$$
$$\dot{u}_i = K_{i-1} u_{i-1} - K_i u_i \qquad [i = 2, \ldots, n+1] \tag{8.20}$$

The linear feedback term

$$A = \left. \frac{\partial G}{\partial u_{n+1}} \right|_{u_{n+1} = 0} \tag{8.21}$$

is applicable *only* in the neighborhood of the stationary state. In the case of negative-feedback systems,

$$A = - \frac{K_0 \left[\sum_{j=1}^{\rho} j\, \alpha_j (x^*_{n+1})^{j-1} \right]}{\left[1 + \sum_{j=1}^{\rho} \alpha_j (x^*_{n+1})^{j} \right]^2} < 0 \tag{8.22}$$

but for positive-feedback systems,

$$A = K_0 \left[\frac{\left[1 + \sum_{j=1}^{\rho} \alpha_j (x^*_{n+1})^{j} \right] \sum_{j=1}^{\rho} j\, \beta_j (x^*_{n+1})^{j-1} - \left[1 + \sum_{j=1}^{\rho} \beta_j (x^*_{n+1})^{j} \right] \sum_{j=1}^{\rho} j\, \alpha_j (x^*_{n+1})^{j-1}}{\left[1 + \sum_{j=1}^{\rho} \alpha_j (x^*_{n+1})^{j} \right]^2} \right] > 0 \tag{8.23}$$

Since x^*_{n+1} is completely defined by ρ, K_0, and the α's, β's, and K's, $A < 0$ for negative-feedback systems, or $A > 0$ for positive-feedback systems, is also defined by these parameters.

The characteristic equation for Eqs. (8.20) is

$$\prod_{i=1}^{n+1} (K_i + \mu) - A \prod_{i=1}^{n} K_i = 0 \tag{8.24}$$

where the μ satisfying Eq. (8.24) are the characteristic exponents. If we set $\mu = 0$, we obtain the function dividing $\mu < 0$ and $\mu > 0$:

$$K_{n+1} = A \tag{8.25}$$

In what follows, we shall refer to Eq. (8.25) as the *equation of zero roots*.

If we set $\mu = i\omega$ [$i = \sqrt{-1}$] we obtain the functions dividing imaginary μ's with positive or negative real parts. These functions for n = 1 through n = 4 are

n = 1:

$$K_1 = -K_2 \tag{8.26}$$

n = 2:

$$\frac{(K_1 + K_2)(K_1 + K_3)(K_2 + K_3)}{K_1K_2} + A = 0 \tag{8.27}$$

n = 3:

$$\frac{(K_3 + K_4)(K_2^2 + K_2K_3 + K_2K_4 + K_3K_4)[K_1^2(K_1 + K_2 + K_3 + K_4) + K_1K_2K_3 + K_1K_2K_4 + K_1K_3K_4 + K_2K_3K_4]}{K_1K_2K_3(K_1 + K_2 + K_3 + K_4)^2} + A = 0 \tag{8.28}$$

n = 4:

$$(B^2 + 4C)^{\frac{1}{2}} - B + 2A = 0 \tag{8.29}$$

$$B = \frac{K_5(2a_0 + a_3^2a_4 - a_2a_3 - 2a_1a_4)}{a_0}$$

$$C = \frac{K_5^2(a_1a_2a_3a_4 - a_0a_3^2a_4 - a_1^2a_4^2 - a_1a_2^2 + a_0a_2a_3 + 2a_0a_1a_4 - a_0^2)}{a_0^2}$$

$$a_0 = K_1K_2K_3K_4K_5$$

$$a_1 = K_1K_2K_3K_4 + K_1K_2K_3K_5 + K_1K_2K_4K_5 + K_1K_3K_4K_5 + K_2K_3K_4K_5$$

$$a_2 = K_1K_2K_3 + K_1K_2K_4 + K_1K_2K_5 + K_1K_3K_4 + K_1K_3K_5 + K_1K_4K_5 + K_2K_3K_4 + K_2K_3K_5 + K_2K_4K_5 + K_3K_4K_5$$

$$a_3 = K_1K_2 + K_1K_3 + K_1K_4 + K_1K_5 + K_2K_3 + K_2K_4 + K_2K_5 + K_3K_4 + K_3K_5 + K_4K_5$$

$$a_4 = K_1 + K_2 + K_3 + K_4 + K_5$$

In what follows, we shall refer to these equations as the *neutrality equations*.

In the parameter space divided by the equation of zero roots and the neutrality equation, the real part of at least one characteristic exponent is negative on one side of the lines and postive on the other. Thus, it is possible to use Eqs. (8.25)-(8.29) to partition the parameter space into areas where at least one characteristic exponent changes sign in its real part. Since unstable singularities are characterized by at least one characteristic exponent with a positive real part (cf. Theorem IV in Chapter 6), this partitioning is equivalent to dividing the parameter space into areas for which the linearized system must be unstable.

In Figs. 8.4 and 8.5 appear the line of zero roots (solid lines) and the neutrality equations (dashed lines) for 2 - 5 vari-

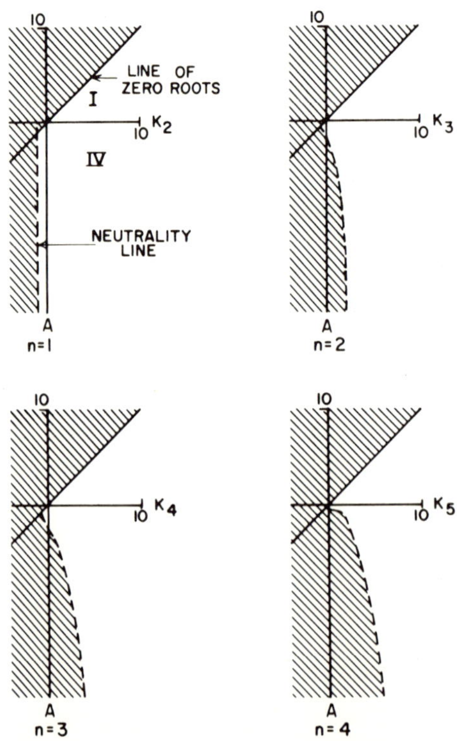

FIG. 8.4. Partition of the linearized feedback term (A) and K_{n+1} for Eqs. (8.20) into areas of stability and instability (slashed areas).

ables in systems described by Eqs. (8.20). In Fig. 8.4, all the $K_i = 1$ except K_{n+1}, which is the parameter used with the linearized feedback term (A) to partition the two-dimensional parameter space. Figure 8.4 illustrates the following about systems described by Eqs. (8.20).

1. Regardless of the number of variables involved, positive-feedback systems (A > 0) possess a constant area of parameter space wherein instabilities must arise. Since K_{n+1} is nonnegative, this area of instability is the shaded area to the left of the line of zero roots in the first quadrant of Fig. 8.4.

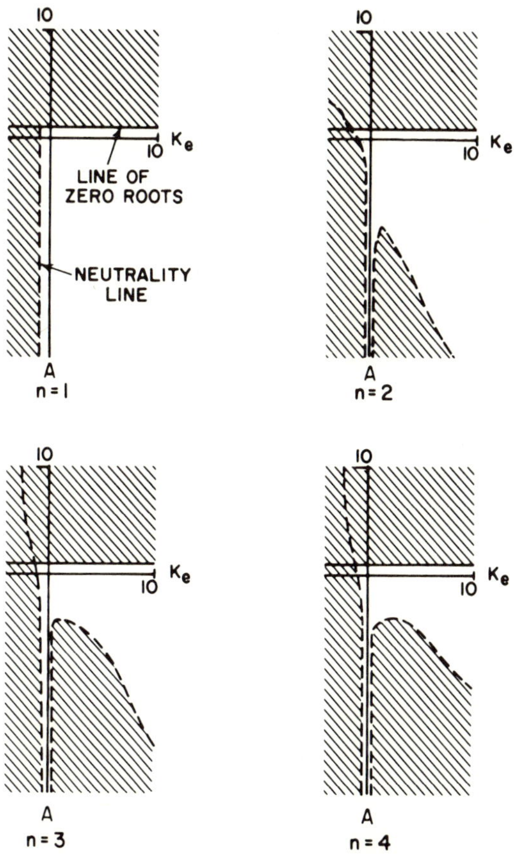

FIG. 8.5. Parition of the linearized feedback term (A) and K_e [e = 1, ..., n] for Eqs. (8.20) into areas of stability and instability (slashed areas).

2. Negative-feedback systems ($A < 0$) involving two variables do not possess any parameter space for $K_2 > 0$ wherein instabilities must arise.

3. Negative-feedback systems involving more than two variables possess an area of parameter space wherein instabilities must arise for $K_{n+1} > 0$; this area increases as the number of variables in the systems increases.

4. Instabilities in both positive- and negative-feedback systems are most likely to occur when K_{n+1} is small. In the case of positive-feedback systems, the relationship between K_{n+1} and A is linear, and it has a slope of unity [cf. Eq. (8.25)]. In the case of negative-feedbcck systems, the relationship between K_{n+1} and A is nonlinear. According to Fig. 8.4, there should be an asymptotic value of K_{n+1} ($K_3 < K_4 < K_5$) above which instabilities need not arise.

In Fig. 8.5 all the $K_i = 1$ except K_e [$e < n$]. K_e is the parameter used with the linearized feedback term to partition the two-dimensional space. Figure 8.5 illustrates that for negative-feedback systems, there is an intermediate value of $K_e = K_i = 1$ at which instabilities must arise for the minimum possible values of $-A$. At values of K_e less than or greater than the other pseudo-first-order constants, a larger value of $-A$ is necessary for instability.

The implication of Figs. 8.4 and 8.5 is that, if stable behavior can be achieved in biochemical systems with negative feedback, it is more likely to be found in systems with weak feedback and large pseudo-first-order constants for the rate of loss of the control chemical. In the case of systems with negative feedback, stable behavior is most likely to be found in systems that also possess a relatively small number of variables and unequal pseudo-first-order rate constants for the enzyme-catalyzed steps. As observed previously, stability is most likely in systems where the control properties are not too strong.

The linearized feedback term for negative-feedback systems wherein the concerted function [Eq. (8.4)] is applicable is

$$A = \left.\frac{\partial G}{\partial u_{n+1}}\right|_{u_{n+1}=0} = -\frac{\rho\alpha K_0 (x^*_{n+1})^{\rho-1}}{[1 + \alpha(x^*_{n+1})^{\rho}]^2} \tag{8.30}$$

Such systems can be partitioned in the two-dimensional parameter space, ρ and $n + 1$ (the stoichiometry of the feedback and the number of variables in the system), into areas of local asymptotic stability and potential instability.[38] The procedure is based on the fact that, since $K_0 > K_{n+1} x^*_{n+1}$, $-A/K_{n+1} < \rho$:

$$0 < -\frac{A}{K_{n+1}} < \frac{\rho\alpha(x^*_{n+1})^{\rho}}{[1 + \alpha(x^*_{n+1})^{\rho}]^2} < \rho \tag{8.31}$$

A root of Eq. (8.31) has a positive real part iff

$$\rho > -\frac{A}{K_{n+1}} > \sec^{n+1} \frac{\pi}{n+1} \tag{8.32}$$

Thus, necessary conditions for instability of the linearized negative feedback systems are met if ρ and $n + 1$ are individually sufficiently large to fulfill the conditions in inequality (8.32). Note that sufficiently large stoichiometries and numbers of variables do not *guarantee* that these systems will be unstable (K_0 or α could be too small, or K_{n+1} could be too large).

Since the secant is never less than unity, instabilities cannot exist for Eqs. (8.20) when $\rho = 1$, no matter how many components exist in the control system. Furthermore, since the secant of $\pi/2$ is infinitely large, there cannot be instabilities when $n + 1 = 2$, no matter how large the stoichiometry of the feedback inhibition is. For all intermediate values of ρ and $n + 1$, the two-dimensional parameter space can be partitioned into areas of potential instability and stability. In Fig. 8.6, the ρ and $n + 1$ parameter space is divided into the section where the stoichiometry and the number of variables are sufficiently large to allow instabil-

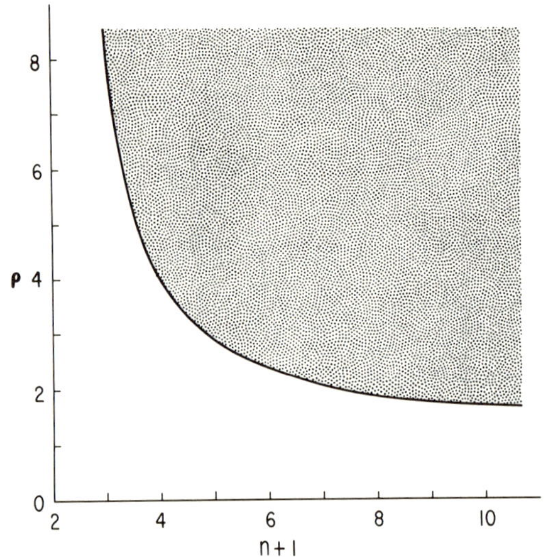

FIG. 8.6. Partition of the parameter space for an enzyme-catalyzed system with negative feedback. The feedback stoichiometry ρ and the number of variables $n + 1$ divide the space into areas of stability and potential instability (slashed areas).

ities to arise (shaded area), and the area where asymptotic stability of the linearized system is guaranteed (unshaded area).

When ρ and $n + 1$ are large enough so that a system falls in the shaded area in Fig. 8.6, it is possible to partition the K_0 and α parameter space into areas of stability and instability. Figure 8.7 illustrates the partition of the parameter space for the flux and the inhibition constant into areas of instability (shaded area) and potential stability (unshaded area) for $K_{n+1} = 1$. In Fig. 8.7 (a), $n + 1 = 3$ (there are three variables), and according to Fig. 8.6 instabilities cannot arise unless the feedback stoichiometry is greater than eight. Therefore, the next higher integer, $\rho = 9$, was chosen to construct Fig. 8.7 (a). The shaded area denotes the values of α and K_0 for which instabilities arise, and the unshaded area represents the values of these parameters when the linearized system can be asymptotically stable. In Fig. 8.7 (b) $n + 1 = 4$ (there are four variables), and instabilities are not possible unless the feedback stoichiometry exceeds four (cf. Fig. 8.6). The next higher integer, $\rho = 5$, was used to construct Fig. 8.7 (b). As

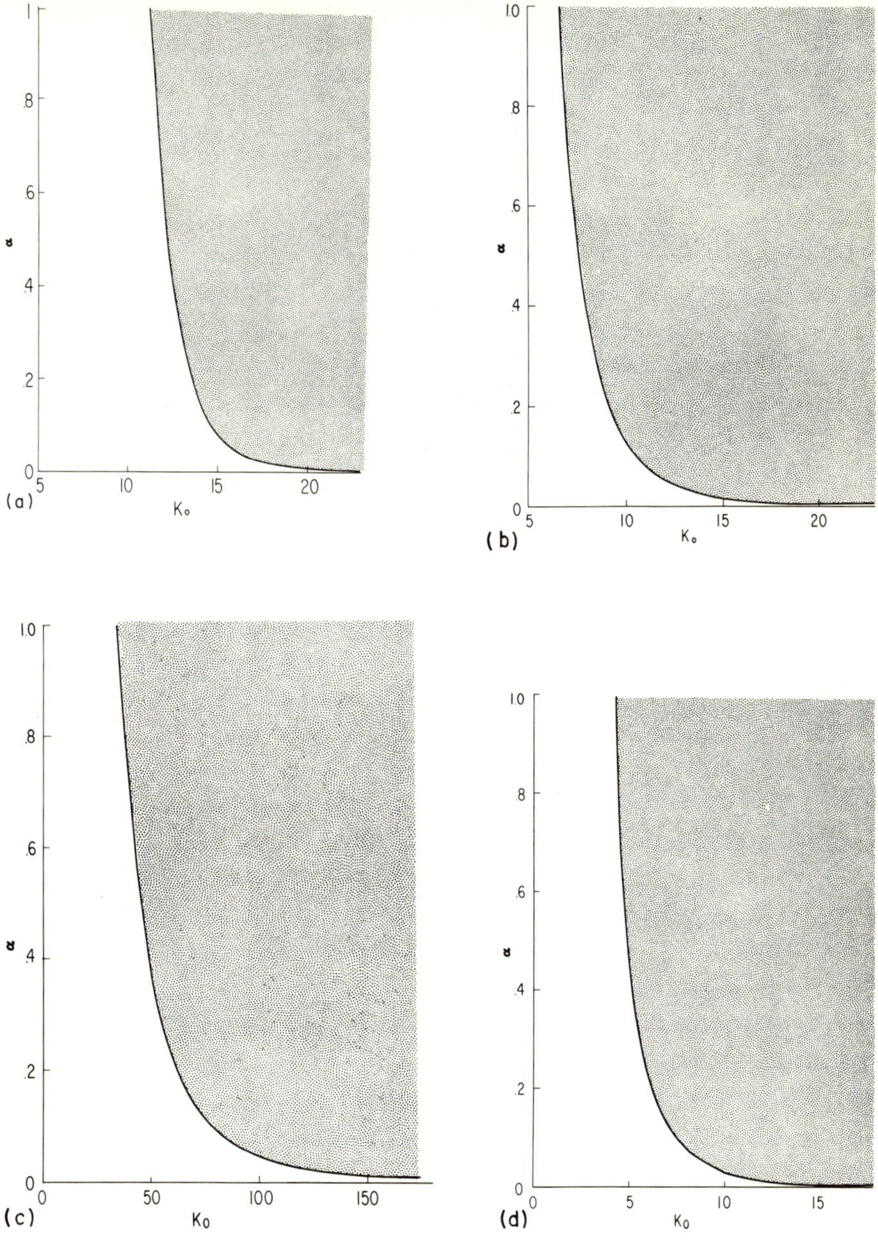

FIG. 8.7. Partition of the parameter space for an enzyme-catalyzed system with negative feedback. The flux K_0 and the inhibition constant α divide the space into areas of stability and instability (slashed areas) for systems involving (a) three variables and $\rho = 9$, (b) four variables and $\rho = 5$, (c) five variables and $\rho = 3$, and (d) five variables and $\rho = 4$.

before, the shaded area denotes the values of K_0 and α where instabilities arise, and the unshaded area indicates the parameter values when the linearized system can be asymptotically stable.

In Figs. 8.7 (c) and (d), $n + 1 = 5$ (there are five variables), and according to Fig. 8.6, instabilities cannot arise unless the feedback stoichiometry is greater than 2.8. In Fig. 8.7 (c), the next higher integer value, $\rho = 3$, was used for the stoichiometry. Note that, because ρ is so close to the minimum value required for instability, K_0 (the flux) can be relatively large without introducing instability [compare with Figs. 8.7 (a) and (b)]. This ability to tolerate larger fluxes without introducing instabilities is unique to the linearized systems [Eqs. (8.20)] involving five or more variables. It is most prominent in systems involving five or eight variables. If only integer values of ρ occur in real biochemical control systems, this property of five- and eight-variable systems could be of selective advantage. For purposes of comparison with the three- and four-variable cases [Figs. 8.7 (a) and (b)], $\rho = 4$ was used to construct Fig. 8.7 (d). As expected, the parameter space wherein instabilities occur is greatest in Fig. 8.7 (d) (five variables), intermediate in Fig. 8.7 (b) (four variables), and smallest in Fig. 8.7 (a) (three variables).

8.5 NONLINEAR ANALYSIS OF BIOCHEMICAL SYSTEMS WITH FEEDBACK CONTROL

In Chapter VII, it was illustrated that the two-component version of Eqs. (8.8),

$$\begin{aligned} \dot{x}_1 &= F(x_2) - f_1(x_1) \\ \dot{x}_2 &= f_1(x_1) - f_2(x_2) \end{aligned} \qquad (8.33)$$

cannot possess limit-cycle behavior as long as the derivatives of f_1 and f_2 are sign definite. Therefore, when f_1 and f_2 are hyperbolic or linear functions, limit-cycle behavior cannot arise for Eqs. (8.33), no matter what the feedback function F is.

The global stability properties of the two- and three-component versions of Eqs. (8.8) with nonlinear negative or positive feedback have been reported[39] for the case when the f's are linear functions:

$$\dot{x}_1 = F(x_2) - K_1x_1$$
$$\dot{x}_2 = K_1x_1 - K_2x_2 \tag{8.34}$$

and

$$\dot{x}_1 = F(x_3) - K_1x_1$$
$$\dot{x}_2 = K_1x_1 - K_2x_2 \tag{8.35}$$
$$\dot{x}_3 = K_2x_2 - K_3x_3$$

In the two-component case, there are two sufficiency criteria for global asymptotic stability of systems with negative feedback:

$$K_1 > 0 \tag{8.36}$$

$$K_1 + K_2 > 0 \tag{8.37}$$

Since the K_1 and K_2 are pseudo-first-order rate constants, they are always nonnegative for all monotonically increasing f. Thus the conditions in inequalities (8.36) and (8.37) are always met, and all negative-feedback systems described by Eqs. (8.34) possess global asymptotic stability.

In the case of the two-component systems with positive feedback, the first sufficiency criterion for global asymptotic stability becomes

$$K_1 < 0 \tag{8.38}$$

Since this condition is not met, one cannot guarantee the stability of two-variable systems with positive feedback by this method.

In the three-component case [Eqs. (8.35)], there are three sufficiency criteria for global asymptotic stability of systems with negative feedback:

$$K_1K_2 > 0 \tag{8.39}$$

$$-a_1 = \frac{K_1}{K_2} a_2 + \frac{K_1}{K_3} a_3 \tag{8.40}$$

satisfied by arbitrary but real a_1, a_2, and a_3, and

$$[K_2(K_1 + K_3) + K_3(K_1 + K_2)]^2 > (K_1 + K_3)(K_1 + K_2)(K_2 + K_3)^2 \tag{8.41}$$

Since K_1 and K_2 are pseudo-first-order rate constants, criterion (8.39) is satisfied, and criterion (8.40) can obviously be satisfied by real a_1, a_2, and a_3. However, criterion (8.41) implies that

$$(K_3 - K_2)[K_2^2(K_1 + K_3) - K_2^2(K_1 + K_2)] > 0 \tag{8.42}$$

If $K_3 > K_2$, inequality (8.42) leads to

$$K_1K_2^2 + K_2^2K_3 > K_1K_3^2 + K_2K_3^2 \tag{8.43}$$

which cannot be true if indeed $K_3 > K_2$. On the other hand, if $K_3 < K_2$, inequality (8.42) leads to

$$K_1K_3^2 + K_2K_3^2 > K_1K_2^2 + K_2^2K_3 \tag{8.44}$$

which cannot be satisfied if $K_3 < K_2$. Finally, if $K_2 = K_3$, the left-hand side of (8.42) is zero, and the inequality cannot be satisfied. This means that one cannot use this method to guarantee the global asymptotic stability of systems with negative feedback and more than two components.

In the case of three-component systems with positive feedback the first sufficiency criterion for global asymptotic stability becomes

$$K_1K_2 < 0 \tag{8.45}$$

Since this condition is not met, one cannot guarantee the stability of three-component systems with positive feedback by this method.

8.6 INSTABILITIES AND THE OCCURRENCE OF LIMIT-CYCLE BEHAVIOR IN BIOCHEMICAL SYSTEMS WITH NEGATIVE-FEEDBACK CONTROL

If a system conforms to Bendixson's[29] negative criterion (cf. Theorem VI), or if the stationary state of a system possesses global asymptotic stability, then the system cannot experience limit-cycle behavior. It is clear, therefore, that autonomous, two-component systems with negative feedback and pseudo-first-order rate laws cannot exhibit sustained concentration oscillations.

The Lyapunov function used to guarantee the global asymptotic stability of the two-component systems cannot be used to guarantee such stability for similar systems involving more than two variables. This could reflect a poor choice of Lyapunov function (perhaps a better choice would have led to a guarantee of stability), or it could be a true indication that under some conditions the multicomponent systems are unstable. The fact that the same Lyapunov function could be used to guarantee the global asymptotic stability of the negative-feedback systems involving two variables, but could not be used to guarantee the stability of identical systems with more variables, suggests that there may be a fundamental difference in the inherent properties of systems with two variables as compared to those of systems with additional variables.[32]

The linear analysis also indicates that negative-feedback systems with two components are stable, while such systems with more than two components can possess unstable singularities. Thus, both the linear analysis and the nonlinear analysis suggests that there may be a fundamental difference in the stability properties of two-component and multicomponent systems with negative feedback.[32]

If negative-feedback systems with more than two variables lack asymptotic stability near the stationary state, the questions arise: What is the fate of the variables in the phase space? Is the system stable but not asymptotically stable: If so, stable limit-cycle behavior around the unstable stationary state is a distinct possibility. On the other hand, can the system be totally unstable? If that is the case, is it stable at the infinities, or do the variables wander aimlessly and chaotically in the phase space? Answers to questions of this type are not easy to obtain. One way to obtain information about the nonlinear behavior is to conduct extensive simulations of the nonlinear equations.

In what follows, we shall consider negative-feedback systems with four, five, and six pseudo-first-order catalyzed steps and $K_0 = 5.1$, $\rho = 4$, $\alpha = 0.08$, $K_i = 1$ $[i = 1, \ldots, n]$, and $K_{n+1} = 0.6$.[40] The characteristic equation for the linearized equations involving four variables is

$$\mu^4 + 3.6\mu^3 + 4.8\mu^2 + 2.8\mu + 2.3277 = 0 \tag{8.46}$$

The four roots of Eq. (8.46) all have negative real parts, so the linearized system is asymptotically stable. Figure 8.8 shows $x_i(t)$ obtained by simulating this system on a digital computer.

The characteristic equation for the linearized equations involving five variables is

$$\mu^5 + 4.6\mu^4 + 8.4\mu^3 + 7.6\mu^2 + 3.4\mu + 2.3277 = 0 \tag{8.47}$$

The five roots of Eqs. (8.47) also all have negative real parts, so the linearized system is again asymptotically stable. Figure 8.9 shows $x_i(t)$ obtained by simulating this system. As in the case involving four variables, the five-variable system possesses global asymptotic stability.

The characteristic equation for the linearized equations involving six variables is

$$\mu^6 + 5.6\mu^5 + 13\mu^4 + 16\mu^3 + 11\mu^2 + 4\mu + 2.3277 = 0 \tag{8.48}$$

Not all of the roots of Eqs. (8.48) have negative real parts, so in this case the singularity is not asymptotically stable. Figure 8.10 shows $x_i(t)$ for this six-variable system. In this case, there are self-sustaining and self-limiting oscillations of the six

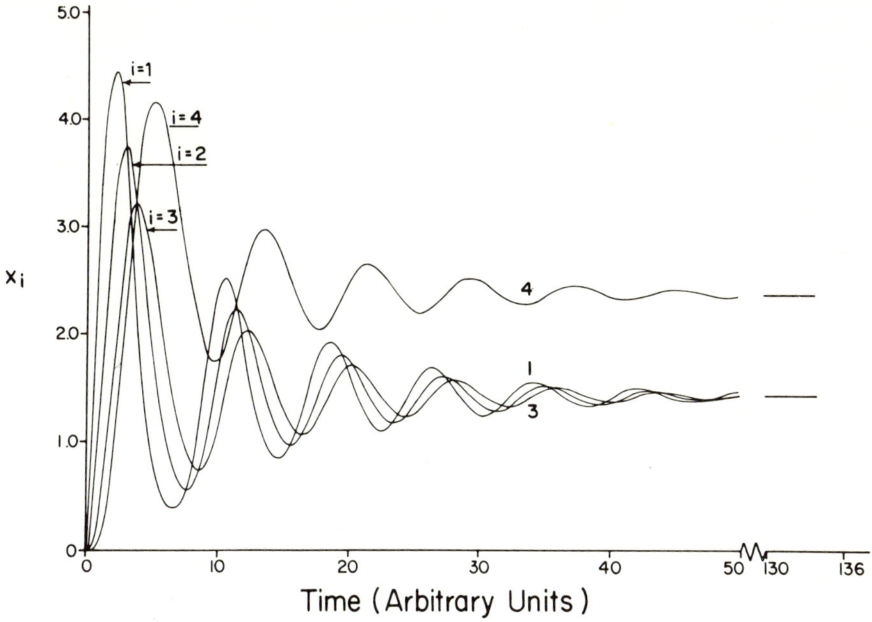

FIG. 8.8. $x_i(t)$ for the components of an enzyme-catalyzed system involving four variables and negative feedback (K_0 = 5.1, ρ = 4, α = 0.08, $K_1 = K_2 = K_3 = 1$, and K_4 = 0.6).

variables. Figure 8.11 illustrates that these oscillations are of the limit-cycle type: If the system is initiated in the neighborhood of the singularity, the variables wind outward to the surrounding limit cycle. If the system is initiated in the phase space outside the limit cycle, the variables wind inward toward the same closed trajectory.

In order to obtain an idea of when limit cycles arise in metabolic systems with negative feedback, simulations of Eqs. (8.4) and (8.8) were carried out on a digital computer for a wide range of n + 1, ρ, K_0, α, K_{n+1}, and rate functions (f). Over 2,500 simulations were obtained[33] when the f were pseudo-first-order rate laws, $f_i(x_i) = K_i x_i = x_i$. These simulations were carried out for all combinations of all possible integer values of ρ = 1 to ρ = 10, three values of α in the range 0.01 - 1.0, three values of K_0 in the range 0.5 - 50.0, and six values of K_{n+1} in the range

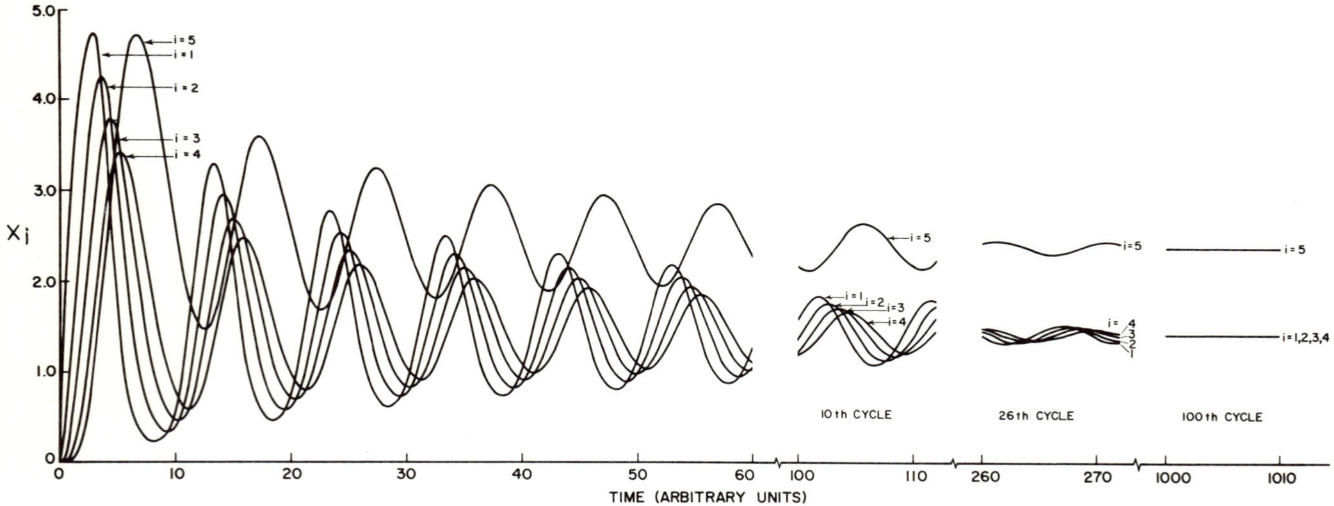

FIG. 8.9. $x_i(t)$ for the components of an enzyme-catalyzed system involving five variables and negative feedback ($K_0 = 5.1$, $\rho = 4$, $\alpha = 0.08$, $K_1 = K_2 = K_3 = K_4 = 1$, and $K_5 = 0.06$).

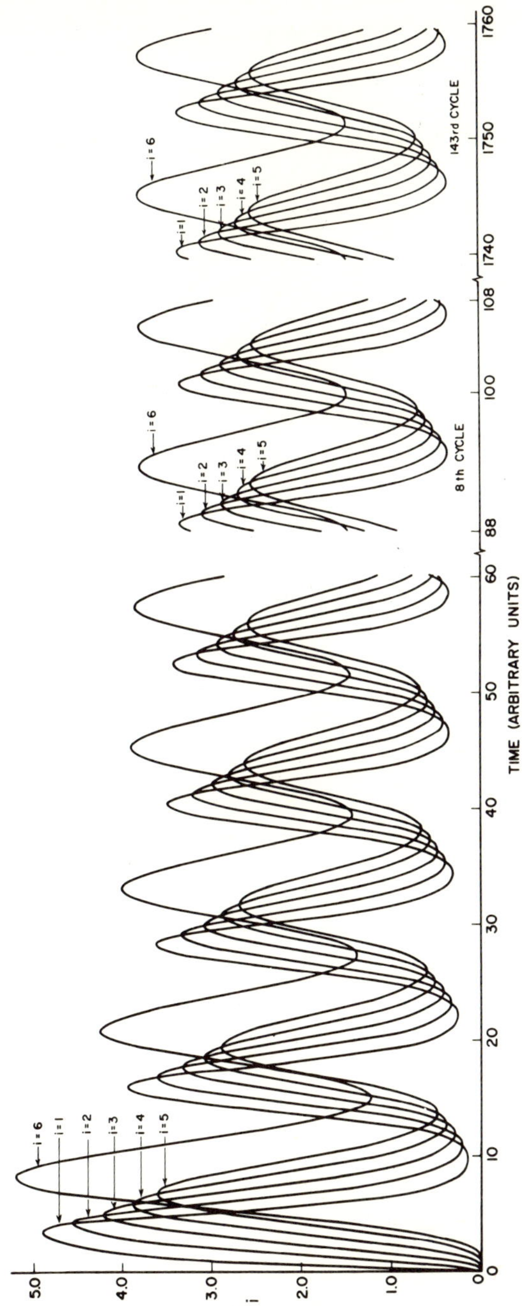

FIG. 8.10. $x_i(t)$ for the components of an enzyme-catalyzed system involving six variables and negative feedback ($K_0 = 5.1$, $\rho = 4$, $\alpha = 0.08$, $K_1 = K_2 = K_3 = K_4 = K_5 = 1$, and $K_6 = 0.6$).

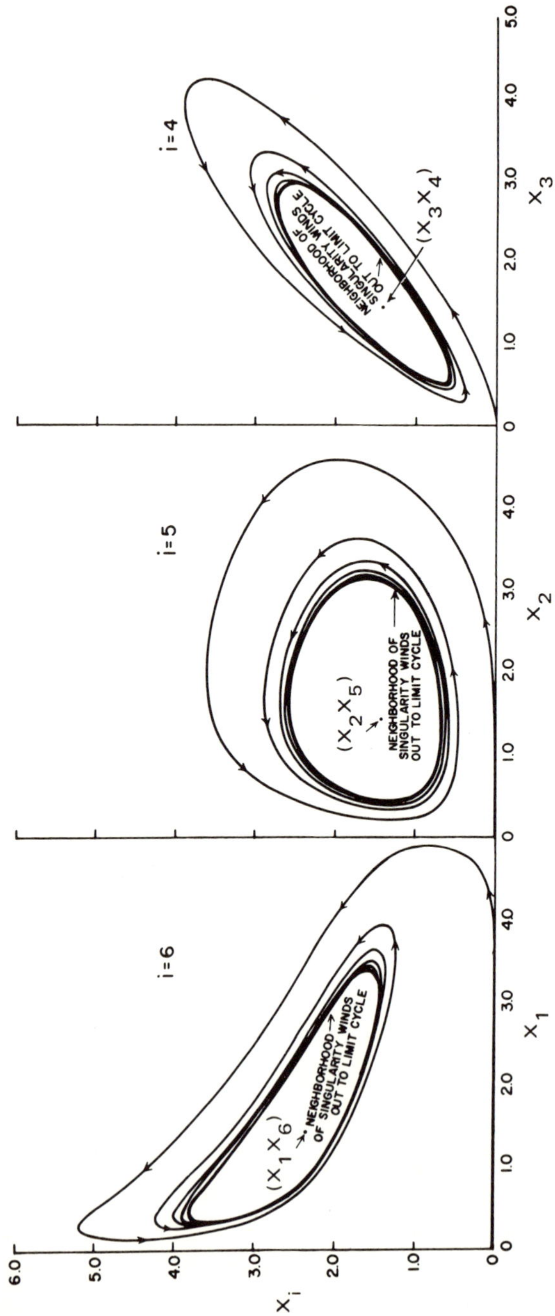

FIG. 8.11. Phase space representations of the six variables in Fig. 8.10.

0.05 - 10.0. The results are summarized in Fig. 8.12. In this figure the squares at the intersection of specific values of K_0 - 0.5, 5.0, or 50.0 (vertical axis) and specific values of K_{n+1} = 0.05, 0.1, 0.5, 1.0, 5.0, and 10.00 (horizontal axis) are filled in if a limit cycle arises. If the intersection is filled in with slashes leaning to the left, the limit cycle was found when $\alpha \geq 1.0$; if the intersection is filled in with diamonds, the limit cycle was found when $\alpha \geq 0.1$; if the intersection is filled in with solid shading, limit cycles were found when $\alpha \geq 0.01$. If the intersection is not filled, the systems involving the indicated values of K_0 and K_{n+1} were simulated for inhibition constants up to unity for each indicated value of ρ and n + 1, but a limit cycle did not occur. For each value of n + 1 in the range of three through eight variables, all 54 possible combinations of the indicated values of K_0, K_{n+1}, and $\alpha = 0.01$, $\alpha = 0.1$, $\alpha = 1.0$ were simulated for all possible integer values of ρ less than those indicated in the figure. At these lower ρ's no limit cycles arose, and Eqs. (8.4) and (8.17) possessed global asymptotic stability.

In each of the simulations carried out, a limit cycle arose when the linearized system was unstable, and no limit cycle arose if the linearized system possessed asymptotic stability. This correlation between local instability and the occurrence of limit cycles is quite striking. Figures 8.4, 8.6, 8.7, and 8.12 illustrate that limit cycles and local instabilities occur (1) at intermediate values of $K_{n+1} = K_i$ for i = 1, ..., n (Figs. 8.4 and 8.12), (2) if the number of variables n + 1 and the feedback stoichiometry ρ are sufficiently high (Figs. 8.6 and 8.12), and (3) if the flux K_0 and the inhibition constant α are sufficiently high. In all cases, limit cycles arose only in nonlinear systems for which the corresponding linearized systems occupy a shaded area in Figs. 8.4, 8.6, and 8.7; when limit cycles did not arise, the nonlinear systems possessed global asymptotic stability, and the corresponding linearized systems occupied an unshaded area in Figs. 8.4, 8.6, and 8.7. In all systems, if a limit cycle or local instability occurs

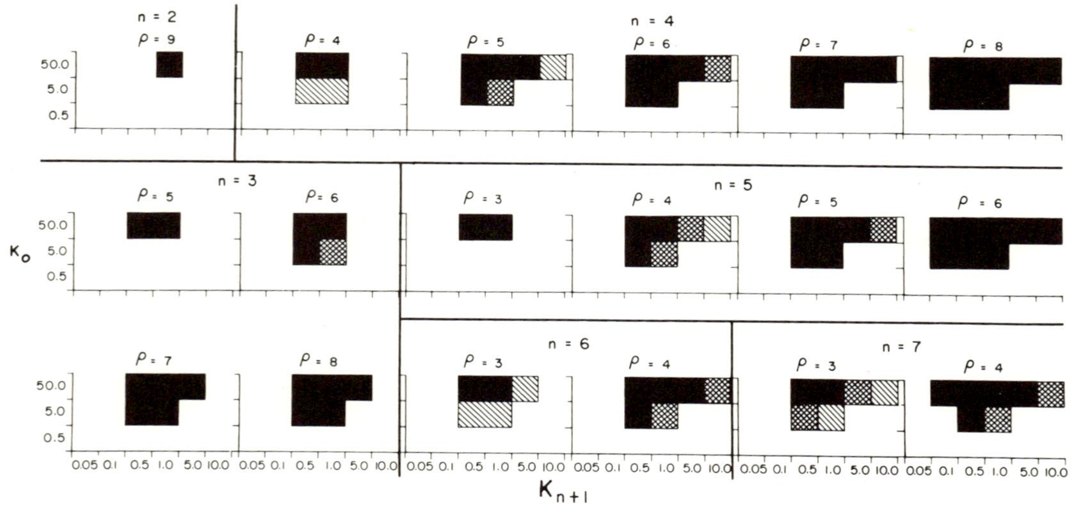

FIG. 8.12. Relationship between the parameter values in an enzyme-catalyzed system with negative feedback [Eqs. (8.4) and (8.17)] and the occurrence of limit-cycle behavior.

for a set of values of $n + 1$, ρ, α, and K_0, then a limit cycle and an unstable singularity will also occur for all larger values of any one of these constants, provided none of the remaining parameters is decreased.

8.7 SENSITIVITY OF METABOLIC LEVELS TO PARAMETER VARIATIONS IN SYSTEMS WITH NEGATIVE FEEDBACK

In controlled metabolic systems with pseudo-first-order rate laws [Eqs. (8.17)] and the concerted negative-feedback function [Eq. (8.4)], the stationary-state level of x^*_{n+1} is the positive real root of

$$\frac{K_0}{\alpha K_{n+1}} - \frac{1}{\alpha} x^*_{n+1} - (x^*_{n+1})^{\rho+1} = 0 \qquad (8.49)$$

The stationary-state level of x^*_{n+1} is most sensitive to parametric variations when the feedback is weak:

$$1 >> \alpha(x^*_{n+1})^{\rho} \qquad (8.50)$$

Substitution of condition (8.50) into Eq. (8.49) leads to the results derived in Sec. 7.5 for metabolic systems without feedback. On the other hand, the stationary-state level of x^*_{n+1} is *least* sensitive to variations in the parameters when the feedback is strong:

$$1 << \alpha(x^*_{n+1})^{\rho} \qquad (8.51)$$

When inequality (8.51) is substituted into Eq. (8.49), we obtain

$$x^*_{n+1} = \left[\frac{K_0}{\alpha K_{n+1}}\right]^{1/(\rho+1)} \qquad (8.52)$$

The maximum controlled sensivity of x^*_{n+1} to changes in K_0 is

$$\frac{\partial x^*_{n+1}}{\partial K_0} = \frac{1}{\rho+1}\left[\frac{1}{K_0}\right]^{\rho/(\rho+1)}\left[\frac{1}{\alpha K_{n+1}}\right]^{1/(\rho+1)} \qquad (8.53)$$

The comparative sensitivities of the controlled stationary-state level [calculated from Eq. (8.53)] divided by the uncontrolled stationary-state level [calculated from Eq. (7.66)] is

$$Z_{K_0} = \frac{1}{\rho+1} \left[\frac{K_{n+1}}{K_0}\right]^{\rho/(\rho+1)} \left[\frac{1}{\alpha}\right]^{1/(\rho+1)} \tag{8.54}$$

Equation (8.54) illustrates the maximum improvement of the insensitivity of x^*_{n+1} to variations of the flux in the controlled system. For a typical set of parameter values (for example, $\rho = 3$, $K_{n+1} = 1$, $K_0 = 5$, $\alpha = 1$), the controlled system is up to fifteen times less sensitive to variations in K_0.

The maximum controlled sensitivity of x^*_{n+1} to changes in K_{n+1} is

$$\frac{\partial x^*_{n+1}}{\partial K_{n+1}} = -\frac{1}{\rho+1} \left[\frac{K_0}{\alpha}\right]^{1/(\rho+1)} \left[\frac{1}{K_{n+1}}\right]^{(\rho+2)/(\rho+1)} \tag{8.55}$$

The comparative sensitivity of the controlled and uncontrolled stationary-state levels is

$$Z_{K_{n+1}} = \frac{1}{\rho+1} \left[\frac{K_{n+1}}{K_0}\right]^{\rho/(\rho+1)} \left[\frac{1}{\alpha}\right]^{1/(\rho+1)} \tag{8.56}$$

Equation (8.56) illustrates that improvement of the sensitivity of x^*_{n+1} to variations of K_{n+1} in the controlled system is the same as the improvement to variations of flux [Eq. (8.54)].

The maximum controlled sensitivity of x^*_{n+1} to changes in α is

$$\frac{\partial x^*_{n+1}}{\partial \alpha} = -\frac{1}{\rho+1} \left[\frac{K_0}{K_{n+1}}\right]^{1/(\rho+1)} \left[\frac{1}{\alpha}\right]^{(\rho+2)/(\rho+1)} \tag{8.57}$$

and the maximum controlled sensitivity of x^*_{n+1} to changes in ρ is

$$\frac{\partial x^*_{n+1}}{\partial \rho} = -\frac{1}{(\rho+1)^2} \left[\frac{K_0}{\alpha K_{n+1}}\right]^{1/(\rho+1)} \ln \frac{K_0}{\alpha K_{n+1}} \tag{8.58}$$

For the typical set of parameter values used previously ($\rho = 3$, $K_{n+1} = 1$, $K_0 = 5$, $\alpha = 1$), the maximum sensitivity with respect to

K_0 is 0.075, the maximum sensitivity with respect to K_{n+1} or α is -0.375, and the maximum sensitivity with respect to ρ is -0.15. Thus, a 10% increase of $K_0 = 5$ to $K_0 = 5.5$ causes up to a 3.75% increase in x^*_{n+1}; a 10% increase of $K_{n+1} = 1$ to $K_{n+1} = 1.1$ causes up to a 3.75% decrease in x^*_{n+1}; a 10% increase of $\alpha = 1$ to $\alpha = 1.1$ also causes up to a 3.75% decrease in x^*_{n+1}; finally, a 10% increase of $\rho = 3$ to $\rho = 3.3$ causes up to a 4.5% decrease in x^*_{n+1}. In the uncontrolled system, a 10% increase of the same values of K_0 or K_{n+1} result in a 50% change in x^*_{n+1}. Note that a change in the number of variables n + 1 does not affect the stationary-state levels of the x* in either the uncontrolled or the controlled systems.

REFERENCES

1. G. E. Briggs and J. B. S. Haldane, *Biochem. J. 19*, 383 (1925).
2. C. F. Walter and M. F. Morales, *J. Biol. Chem. 239*, 1277 (1964).
3. C. F. Walter, *J. Theor. Biol. 15*, 1 (1966).
4. C. F. Walter, *J. Theor. Bid. 44*, 1 (1974).
5. E. L. King and C. Altman, *J. Phys. Chem. 60*, 1375 (1956).
6. D. D. Fisher and A. R. Schultz, *Math. Biosci. 4*, 189 (1969).
7. A. R. Schultz and D. D. Fisher, *Canadian J. Biol. Chem. 47*, 889 (1969).
8. K. E. Iverson, "A Programming Language," Wiley, New York, 1962.
9. K. E. Iverson, *Comm. Assoc. Comp. Mach. 7*, 80 (1963).
10. K. E. Iverson, *IBM Systems J. 2*, 117 (1964).
11. M. J. Barrett and C. F. Walter, *Enzymologia 38*, 10 (1970).
12. G. C. Nooney, *J. Theor. Biol. 11*, 46 (1966).
13. E. M. Chance, E. P. Shephard, and A. R. Curtis, "Handbook for Use of the University College London Enzyme Simulator and Optimization," London, 1972.
14. B. Woolf, quoted in J. B. S. Haldane and K. G. Stern,

"Allgemeine Chemie der Enzyme" Steinkopff-Verlag, Dresden and Leipzig, 1932, p. 119.

15. A. Walker and C. Schmidt, *Arch. Biochem. Biophys. 5*, 445 (1944).

16. G. A. Fleisher, *J. Am. Chem. Soc. 75*, 4487 (1953).

17. B. H. J. Hofstee, *Science 116*, 329 (1952).

18. C. F. Walter, "Steady-state Applications in Enzyme Kinetics," Ronald Press, New York, 1965, Chapter 5.

19. C. F. Walter, *J. Biol. Chem. 249*, 699 (1974).

20. C. F. Walter, *Biochemistry 1*, 652 (1962).

21. C. F. Walter and M. J. Barrett, *Enzymologia 38*, 147 (1970).

22. M. J. Barrett and C. F. Walter, *Enzymologia 38*, 161 (1970).

23. F. G. Heineken, H. M. Tsuchiya, and R. Aris, *Math. Biosci. 1*, 115 (1967).

24. N. Minorsky, "Nonlinear Oscillations," Van Nostrand, Princeton, N.J., 1962.

25. E. A. Barashin and N. N. Krasovsky, *Dokl. Akad. Nauk USSR 36*, 3 (1953).

26. P. R. Halmos, "Measure Theory," Princeton University Press, Princeton, N.J., 1964.

27. H. Poincaré, *J. Math. 7*, 3 (1881).

28. C. F. Walter, in "Biochemical Regulatory Mechanisms in Eukaryotic Cells," E. Kun and S. Grisola, eds., Wiley Interscience, New York, 1972, Chapter 11.

29. I. Bendixson, *Acta Math. 24*, (1901).

30. J. Hearon, *Bull. Math. Biophys. 15*, 121 (1953).

31. L. Onsager, *Phys. Rev. 37*, 405 (1931).

32. C. F. Walter, in Springer Lecture Notes in Mathematics, " "Nonlinear Problems in the Physical Sciences and Biology,"

I. Stakgold, D. Joseph, and D. Sattinger, eds., Springer-Verlag, Berlin, 1973, p. 343.

33. C. F. Walter, *J. Theor. Biol. 27*, 259 (1970).

34. E. Stadtman, First Symposium on Control of Metabolism, 1969.

35. C. F. Walter, *J. Theor. Biol 23*, 23 (1969).

36. R. Ruffini, "Riflessioni Intorno Soluzione delle Equazioni Algebraiche Generali," Modena, 1813.

37. C. F. Walter, *Comput. Biol. Med. 1*, 215 (1971).

38. G. Viniegra-Gonzalez and H. Martinez, *Proc. Biophys. Soc. 13*, A-210 (1969).

39. C. F. Walter, *J. Theor. Biol. 23*, 37 (1969).

40. C. F. Walter, *Biophys. J. 9*, 863 (1969).

INDEX